HAWKS IN FOCUS

A STUDY OF AUSTRALIA'S BIRDS OF PREY

HAWKS IN FOCUS

A STUDY OF AUSTRALIA'S BIRDS OF PREY

JACK CUPPER
LINDSAY CUPPER

JACLIN ENTERPRISES

MILDURA ● AUSTRALIA

First Published 1981
by
JACLIN ENTERPRISES
MILDURA ● AUSTRALIA

ISBN 0 9593975 0 7

Wholly Designed, Set-up and Printed
by
T. & V. JAMES OFFSET PRINTING
MILDURA ● VICTORIA ● AUSTRALIA

Laser Scanned Positives by ENTICOTT/POLYGRAPH, MELBOURNE

FOREWORD

What would motivate a middle aged farmer, also known as a 'hard-headed' business man in his local community, suddenly to cut right back on these and associated activities, and devote his active mind and impressive physique to the gentle art of bird watching?

The short answer could be, a minor heart attack. But to say that was the beginning of Jack Cupper's interest, in both birds and photography, would be an over-simplification. Both date back to boyhood.

During the difficult years usually known as the 'threadbare thirties' he spent much of his time in the far outback of New South Wales, working at a wide range of bush jobs, living close to nature, to whose charms few men are completely insensitive.

The 1939-45 war took him far away. Serving with the 2/7th Infantry Battalion of the A.I.F., he lost an eye while soldiering at Tobruk.

After repatriation and demobilisation Jack spent many creative years in the district around Mildura, which has become famous for its vines and citrus and the products thereof. Always it was the challenge that attracted him; as soon as he had solved the problem of turning hitherto almost worthless land into prime orchards and vineyards he began seeking fresh fields to conquer. For a while, the mediocre performance of the administrative and marketing branches of his own industry attracted his attention, and he soon made his voice heard and heeded among top management.

Then illness struck, and confronted him with the possibility of becoming the wealthiest man in the cemetery, in the not too distant future. Obviously, physical and mental pressures in the farming and business spheres had to be abated.

Historically, many well known soldiers have turned to ornithology, so different in every way from the disciplines imposed by their profession, but demanding keen perception, a high degree of physical fitness and an analytical mind.

Jack soon perceived a gap in the recorded knowledge of birds. For a variety of reasons, hawks and eagles have attracted a disproportionate share of man's attention since the dawn of history. But relatively little was known of the intimate details of their family lives, especially at nesting time. The majority of Australian raptors had seldom, if ever, been photographed at the nest. Living in remote areas and nesting in inaccessibly high trees, they had hitherto defied the skill of photographers to capture their home life on film.

Here then was a real challenge. Even to locate all Australian species was a herculean task. But Jack and his son Lindsay had motivation aplenty, and they set off with gusto to write a new chapter in the literature of ornithology. Lindsay's youthful enthusiasm and vigour, his skill with cameras, and his willingness to endure considerable physical hardship made the whole project seem feasible.

How well these men succeeded, each reader may judge from the ensuing pages. Their words and pictures need little commentary; they speak for themselves.

I believe ornithologists everywhere, whether amateur or professional, full time or part time, and indeed anyone with an interest in the beauty and complexity of birds in their natural environment, will welcome and find delight in this original, factual and exciting book.

A. C. Cameron
"Rockwood", Chinchilla, Q. 14/5/81

CONTENTS

INTRODUCTION

In recent years there has been a great change in the general attitude of people toward birds of prey. Not so long ago many were considered vermin and one species had a bounty on its head in many areas of the country. Other species, notably Goshawks, although not having bounties paid for their destruction, had been declared pest fauna as recently as 1952 in Queensland and were not removed from that list until June, 1971. Probably no raptor anywhere in the world has suffered such an onslaught as our Wedge-tailed Eagle.

Countless thousands were destroyed each year with little consideration for their ecological role. Today their importance - especially in the control of rabbits - is gradually being recognised and most states have enacted legislation for their protection. Governments however, cannot be blamed for the earlier destruction, they generally reflect the attitudes of the people of the day. Farmers and graziers whom we have met in recent years agree that this raptor should be protected. Many other raptor species have suffered unjustly at the hands of man and only the vastness of our country and its relative inaccessibility has previously afforded them some protection. The one time Australian attitude to one's surroundings, sometimes expressed as, "If it moves, shoot it. If it doesn't cut it down", seems to now be giving way to a more enlightened attitude.

We, (father and son), have always been interested in birds, although birds of prey were our first love. By 'Birds of Prey' we mean all diurnal raptors: eagles, harriers, kites, falcons and goshawks. The decision to attempt a photographic study at the nest of all twenty four species found on our continent, was taken in 1974. Initially it was to be a form of relaxation, a hobby, which would probably keep us amused for a year or two. Before we had finished the first season our hobby had become almost an obsession, occupying the mind most of the day and much of the night, working out ways to improve our approach to the project. By that time the idea of publishing a book was well established.

Much of the literature published on Australia's birds of prey in recent years has been a repetition of findings published many years ago. There was little that could be regarded as new, indicating a study of the literature rather than the birds themselves. Many of the publications depicted paintings of their subjects, some no doubt taken from museum skins and incorporating a certain amount of artistic imagination that could often make them useless as field guides. Some authors had used caged, pet, trapped or tied-down birds for their photographic studies. One, perhaps to his credit, admitted he wasn't a purist. We have no argument with them, as they were in the best position to assess their own limitations. That type of photography presented very little challenge for us and without a challenge there isn't the same sense of achievement. A publication that we admired and gave us a lot of inspiration was Peter Steyne's Eagle Days. The fine photography of his subjects, African eagles, all taken in the wild, was the photographic standard we set ourselves to emulate.

All photographs are our own as this was part of the challenge that made the project so attractive. We knew little about photography when we started and realise now that we should have had some preparatory tuition in the art. It would have been far more economical from a financial point of view and certainly less frustrating when we compared our earlier efforts with those of others.

We lay no claims to being experts on birds of prey, but naturally our close study of them has taught us much and we have been able to shed further light on the habits of some of them. Several had never been studied to any great extent and some not at all, so we had the field almost exclusively to ourselves. Some remarkable behaviour previously unheard of has been recorded herein and supported photographically. We observed a unique occurrence of inter-breeding between Brown and White Goshawks and a strange relationship between our third largest raptor, the Black-breasted Buzzard, and our smallest, the Australian Kestrel.

Our story should also be of some assistance to potential bird photographers as we have gone into some detail throughout to explain our methods of operation, equipment used and the areas in which we operated - including a little of their history, derived on occasions from the inhabitants.

In order to make this book more enjoyable for the general public the text is in the vernacular, and scientific names have been omitted. For serious ornithologists a glossary of scientific names appears at the back of the book. All English names conform with the Recommended English Names for Australian Birds (EMU. Vol. 77 May 1978).

At the end of each chapter dealing with individual species we have included a brief summary of the information available in previous publications, supplemented and annotated with our own observations, where appropriate.

An early problem was to devise a means whereby we could obtain our photographs safely without the risk of having the adult birds desert their nest, as the welfare of the birds was always our prime consideration. We tried several concepts, but they proved too hazardous, unwieldly, or time consuming. Through trial and error we gradually developed our present set-up: basically an aluminium telescopic tower of up to five inter-locking sections. It has proven a tremendous asset as it has allowed us to work several species at the nest for the first time.

The use of a tower is by no means a new con-cept, but the type used, incorporating so many refinements for speed of erection, portability, safety and great scope in respect to height attain-able, has enabled us to cover in a few seasons what could not otherwise have been accomplished. We usually set-up the tower well back from the nest and do not raise it to the full height required until the birds have become used to it, or at least accepted it in that position. Once accepted there, it is moved into the desired position, which may be anything from four to ten metres from the nest, and gradually raised to nest height. The distance from which we work is governed by several factors, the raptor's size, its general known behaviour, and the focal length of the lens we decide to use. The reaction of raptors even within the same species is not always the same. Acting with due caution we soon learned to overcome any inhibitions in the bird's behaviour.

As we are active fruit-growers, the time we could spend away from our properties was often very limited. We travelled in excess of 350,000 kilometres in the seven years it took to accomplish our task. There were several trips to cover many of the species at various stages of their develop-ment and also to upgrade some of our earlier work. Some trips were abortive as we failed to find our quarry and others were aborted through weather conditions.

1974 proved to be a good year to start our pro-ject. The three wet years from 1973 to 1975 trans-formed the inland. The average annual rainfall for the Innamincka - Birdsville area is about 150mm (6 inches). In 1974 there was almost 1000mm (40 inches), and there were 500mm in 1975. The native rat was in plague numbers along the Birdsville Track and provided ample food for many thousands of raptors. Rabbits, introduced from Britain in 1859 and now established over a large portion of Australia, were also in plague numbers along the Strzelecki Creek. It was in these two areas that we did much of our work from 1974 to 1979

Driving only conventional drive vehicles during the first four years of the project caused us some problems at times on wet and flooded roads. When travelling in good conditions or on sealed roads we often did tremendous mileage per day. By starting at 03:30 and maintaining a steady 100-110 kilometres per hour we usually managed at least 1500 kilometres for the first day. There is little joy in that type of travelling, but to get our work done in the time available there was no

alternative. We ate as we drove and snatched a few hours sleep on the roadside - in our swags under a tarpaulin when wet, but when warm and dry just "veiled our eyes from moon and stars" and slept on the ground, often in full marching order.

We've sat for countless hours in a tiny hide at heights up to thirty metres in extremes of heat or cold, where one could shed a kilogram or more a day in a 'Turkish Bath' or conversely sit with feet numb from the bitterly cold conditions and wonder how those steps down to the ground are going to be negotiated on feet that have no feeling. When having sat for eight or more hours without any appreciable movement, muscles in legs, back and buttocks, singly, and at times in unison scream their protest. Life wasn't easy, whether it was meant to be or not, but the harder we worked the luckier we appeared to be. Often, as we sped through the night, ever alert for kangaroos or straying cattle on the roads, we wondered was there anyone, anywhere in Australia, as hell-bent on a task as us?

It has taken us seven seasons (usually from June to January), on a part-time basis, to collect the photographs and data for this book. Reaching that goal has given us much personal satisfaction as well as a wealth of knowledge obtained first hand - some of it, we believe, previously unknown and certainly not recorded. But the adage, "It's the journey, not the goal, that matters", has certainly been true for us. The kindness, assistance and hospitality of people we met along the way made our quest easier, faster and most enjoyable. With few exceptions farmers, graziers, photo-graphers and people in general co-operated with us everywhere we went. We have published this book to share with them, the public in general and bird-watchers in particular some of the beauty of our birds of prey in the wild. If it does anything towards the conservation of these birds, then the rewards are manifold.

It is a pleasure, after the seven years that it has taken to complete this project, to be able to acknowledge the generosity of numerous people.

We are deeply grateful to A. C. (Cec) Cameron, M.B.E., D.F.M., and his wife Jean, of 'Rockwood' via Chinchilla, Queensland. Cec's constructive criticisms, suggestions and assistance were invaluable both in the field and in the preparation of the manuscript for this book.

We are also most grateful to Drs. David and Margaret Hollands of Orbost, Victoria. David had been photographing birds for many years and we learned much from him. He had a similar interest to our own in birds of prey and often worked with us. We were able to reciprocate by sharing our towers at many of the nest-sites. On one memor-able occasion David climbed twenty three metres up an unsteady tower to deliver with trembling hands a cut-lunch to the occupant of the hide - who had by then completely lost his appetite.

There were many others who contributed, of course, from Wenlock in the far north of Queensland to Port Lincoln in the south of South Australia and a hundred places between. We are grateful to the Gillison family of Louisa Creek, via Mackay, Queensland, and to the brothers Ted and George Riek and their families of 'Merty Merty' and 'Bollard's Lagoon' Stations in the north-east of South Australia, for the hospitality for which the people of the outback are renowned.

Being farmers as well as having other business interests it wasn't always possible for both of us to get away together. Travelling in isolated inland areas where telephones are non-existent, the only communication being through the Flying Doctor Radio and station homesteads over a hundred kilometres apart, it was essential that we didn't travel alone. In the first years of the project Albert Chamberlain or Jim Hogg provided companionship and assistance for one of us when the other couldn't get away. Besides undertaking the chores connected with camping in the bush it is essential for one to be 'seen into the hide'. The theory is that birds can't count so if two people walk up to a hide, one enters and the other departs, the watching birds will accept the hide as being vacant. (We have reservations about birds being unable to count, but more on that later). We are grateful to Albert and Jim for their assistance.

We sought and received information from the Curators of Birds at several museums and from the staff of the Wildlife Division of the Commonwealth Scientific and Industrial Research Organisation (C.S.I.R.O.) Canberra. At the latter Mrs. Billie Gill 'adopted us', as she has so many others, and gave us valuable assistance. Articles from the 'Emu', appropriate to our project, were selected and forwarded to us by Mrs. Joan Vincent of Bairnsdale, Victoria, and by David Baker-Gabb of Monash University, Melbourne. Ron Taylor of Balaklava, South Australia, was responsible for sending copies of the 'South Australian Ornithologist' and constructively criticising the manuscript. Norman J. Favaloro of Mildura, Victoria, made his extensive library and egg collection available to us.

Finally, without the help of our respective wives - in looking after our business interests while we were away, in stoically bearing with our early morning departures and arrivals, and in tolerating this seven year obsession - this book would have been well-nigh impossible.

May all others who helped by deed or word find a little extra enjoyment from this book in the knowledge that they, too, contributed.

Jack and Lindsay Cupper.

EQUIPMENT

Because we began our project with little idea of the demands it would make, both on our time and our resources, we developed or acquired our equipment as the need arose. Every situation presented problems which were only solved by trial and error until we arrived at our present stage of operation. No doubt we will continue to refine our methods and improve our equipment in the future.

During the first four years we used Ford Falcon utilities for the transportation of ourselves, towers and equipment. In the fifth year we acquired diesel-powered four-wheel-drive vehicles so that the more rugged and remote areas of the country could be searched with a greater degree of confidence. A Honda All-Terrain Cycle, fitted with balloon tyres proved a boon when searching for nests; on it as much country could be covered in a day as would take a week on foot. Along navigable waterways and on lakes we used a small aluminium boat with an outboard motor.

We designed and built our towers ourselves. They consist of five aluminium telescopic sections held together by over two thousand pop rivets plus a gross or more of bolts where they could be used without fouling. The sections were constructed with a mere 3mm clearance between each, so even a slight distortion can render them inoperable. Many weeks of work can be wiped out by a storm if the towers are not properly secured by guy ropes. Various locking devices ensure each section is securely locked in position. If a section above slid down while one was climbing, fingers and toes would be neatly guillotined - at least a life of disfigurement would be saved because the resulting fall would be fatal. A five section tower allows us to work nests up to thirty metres and weighs approximately 275 kilograms. We built five of them at a cost of $3,000 each.

The towers were transported on a specially designed and built steel frame mounted onto our vehicle. At our home base they were loaded by forklift. At sub-bases in Queensland and South Australia we used manpower. During the breeding season they were usually standing at nest sites and when one was required elsewhere it was wound down, the base pegged so it could be lowered by leaning against the carrier frame on the vehicle, and the latter driven forward to the tower's centre of balance. The pegs securing the base were removed and the tower was pushed horizontally onto the carrier. Four bolts held it in place during transportation.

The erection of the tower followed the reverse procedure. The base was lowered to the ground, pegged and the vehicle reversed until the tower

was vertical. Guy ropes were attached to each section. The sections were elevated to the required height by hand winches built into the base. As each section was raised the guy ropes were tied firmly to nearby trees, or to steel pegs hammered into the ground. A minimum of four ropes were used on each section, although the towers were constructed with provision for eight. After having towers damaged in gales and cyclones we usually used extra guy ropes if we anticipated leaving them in place for more than a day or two. Normally two men would erect the tower, but it was possible for one to do so.

Having several towers allowed the luxury of setting them up at various nests some days prior to working them. Since it was important the birds behaved naturally, we usually gave them a few days to accept this newcomer in their immediate surroundings. There is little value in a cautious transfixed bird staring at the camera.

The collapsible canvas-covered hide is framed with aluminium and measures 0.9 metres long, 1.2 metres high and 0.75 metres wide (3' x 4' x 2'6''), with a 3mm thick sheet aluminium floor. Although a larger hide would certainly be more comfortable, we found these dimensions were necessary for minimum wind resistance, thus lowering the

4

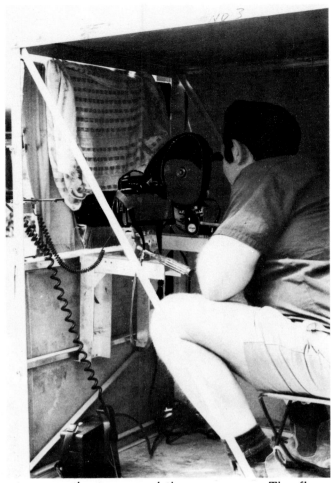

stress on the tower and the guy ropes. The floor was carpeted to muffle the sounds of feet and equipment movement, and provided some insulation for the feet in cold weather. We usually sat on padded camera cases or small fold-up canvas stools. Instead of using tripods, which constrict feet and leg room, we hung adjustable stands with camera mounts onto the frame. These were adapted from an idea developed by a friend, the late Ron Taylor, who made mounts to secure his camera when filming from his vehicle. They were made from aluminium angle sections and could be adjusted both vertically and horizontally. We mounted two cameras, a Beaulieu R16 and a Mamiya RB 67, side by side on individual stands. The top half of the hide was curtained so only the camera lenses protruded, and a towel or discarded clothing was stuffed in the slits in the curtains above and below the lens. On the outside of the hide we fitted Braun F900 (later, Braun F910) flash heads - using just one head unless working close, when a modelling effect could be achieved - and ran the leads to a battery on the floor.

We used a number of cameras and accessories for our still photography. We began with two 35mm cameras: a Minolta SRT101 with TTL metering and 300mm f/4.5 Rokkor lens, and an Ashai Pentax Spotmatic SP11 with TTL metering and 300mm f/4 Takumar lens. The Rokkor focussed down to 4.5 metres and the Takumar 5.4 metres. Later these cameras were mainly used for close-up shots of eggs and chicks when the Minolta, in particular, was equipped with either a 50mm or a 28mm lens. A 6 x 7 Pentax with TTL metering and 105mm f/2.4 and 300mm f/4 Takumar lenses was excellent for flight, habitat and high speed landing shots. We purchased a Novaflex follow-focus 400mm f/5.6 lens for flight shots on the 35mm Pentax, but we never really mastered the art of using it. For these shots we tended to use either the 6 x 7 Pentax, which gave us good or better results, or a Minolta XD 7 with motor drive up to 2 f.p.s. coupled to the 300mm f/4.5 Rokkor. We would have liked to use the XD 7 for action at the nest, but chose not to for several reasons: the hide was always too close for the 300mm Rokkor, we had no suitable replacement lens and, in any event, there was never enough time to switch lenses.

Our main cameras for still photography were two Mamiya RB 67 Pro-S's with TTL metering and the choice of three Mamiya-Sekor long lenses all with built-in leaf shutters: 90mm f/3.8 for mostly scenic shots, 250mm f/4.5 with a minimum focus just under two metres for work from two to three metres from the nest, and 360mm f/6.3 with a minimum focus just under four metres. The 360mm was also used with a No. 1 extension tube which lowered its minimum focus to about two metres. We chose between two focussing screens- split image and fresnal prism - when the need arose. We found that the 6 x 7 detachable film packs for the RB 67s were invaluable for switching between different-rated film, replacing exposed packs and for either horizontal or vertical format. Kodak film was used almost exclusively - beginning with Ektachrome 64 ASA and 160 ASA, then EPR 64 and EPD 200 when it became available.

Both 6 x 7 cameras, the Pentax and the Mamiya, had their advantages and their disadvantages. The Pentax had greater speed, up to 1/1000 second, although for flash it only synchronized to 1/30 second. It needed only one movement to advance the film and cock the shutter. Altering the focus in the hide wasn't as simple as it was with the Mamiya equipped with bellows; it required extending the hand along the lens barrel or drawing the camera into the hide - a painstaking job with a wary bird. Reloading, furthermore, wasn't as convenient as with the Mamiya, nor could the film be switched. The Mamiya could be synchronized for flash up to 1/400 second (although this was still too slow to stop such action as high-speed landing), had a choice of focussing screens, and interchangeable film packs. It had one major drawback: we couldn't take quick shots in succession because advancing the film and cocking the mirror required two movements.

It is important to note that the use of any SLR can result in blurred photographs. The sound of the mirror moving up can cause the bird to react and, although the delay between that and the shutter operation is only a fraction of a second, the photograph can be consequently unsatisfactory. One way toward avoiding this problem is to photograph with the mirror up. This isn't always successful either - the bird may move out of focus, or may not alight in the pre-framed and pre-focussed area.

For filming we used two Beaulieu R16's using automatic exposure and powered by rechargeable batteries: in the bush this was done with a 240 volt generator. These versatile cameras were equipped with 12-120mm f/2.2 Angenieux zoom lenses, although an adaptor ring enables us to fit an 85-210mm Takumar zoom when necessary. We used the fine-grained Eastman Colour Negative 11 Film 7247 rated at 64 ASA with a Wratten 85 Filter for daylight and 100 ASA without the filter at night. It was supplied on 100 foot spools. We tried magazines holding 200 feet of film and found them unwieldy for storing when they were on the camera; it was not always convenient to use all the 200 feet before the magazine was detached. When filming at night we lit the nest with two 1000-watt flood-lights supplied by a 2500-watt generator driven by a petrol engine. We attempted to use a Uher 1200 Report Syncro recorder for synchronized sound-recording at the nest, but we found we could not blimp the Beaulieu effectively and so we didn't use it as much as we would have liked. AT home a Minette Viewer and a Siemens projector enabled us to study the results of our efforts.

We never carried tents because erecting and dismantling them wasted precious time. Instead we carried tarpaulins and used them when it rained. A small three-way refrigerator (12-volt battery, 230-volt AC, or LP gas) proved remarkably versatile and indispensable, protecting food and film from the heat.

It wasn't possible to transport all this equipment on one vehicle at one time. Since our towers were our largest and heaviest item we stored them at strategic points around the country. This tactic saved considerably in both fuel and time when travelling to our locations.

BLACK-SHOULDERED KITE
Elanus notatus

This extremely attractive small Kite favours open grasslands, woodlands and cultivated areas throughout Australia, usually in the coastal regions. Once it was thought not to be found more than five hundred kilometres from the coast. However, the Kite is nomadic, its movements governed by the availability of prey. When conditions have been suitable we have seen one or two pairs around the edge of the Simpson Desert and other far inland areas. In the northwest of Victoria its population is variable. Every year from autumn through until spring small numbers arrive and breed around the irrigation districts near our home. In some years, when food is plentiful, large numbers arrive and breed in the drier outlying wheat country.

The unusually wet months of 1974 provided ideal conditions with plenty of mice and rabbit kittens. Within a two-hour drive of our home we found ten pairs nesting in Mallee trees along the tracks between wheat farms. At one nest the Kites appeared to feed their four chicks almost entirely on mice until three of them were flying. While the nest containing the remaining chick was under observation two of the flying juveniles returned and solicited food. A moment later the female Kite crashed onto the rear of the nest and with much wing-beating struggled into it carrying a three-quarter-grown rabbit, a great effort for such a small bird. Normally prey consists of mice and insects, although we recorded lizards, frogs and rabbit kittens. Most prey is taken on the ground. When hunting the Kite hovers with its body hanging almost vertically as if suspended from the stationary head. There is no apparent lateral movement at all, even in a stiff breeze. This action differs markedly from that other superb exponent of hovering, the Australian Kestrel.

The female Black-shouldered Kite looks on as her flying chicks feed on a rabbit.

The Black-shouldered Kite in flight.

The female Black-shouldered Kite broods her tiny chicks.

The following year we were unable to locate any nests, although there were a few pairs of kites about. In 1976 there was a locust plague coinciding with an increase in the population of mice and once again we found the kites breeding.

In mid-December we inspected a nest containing two freshly hatched chicks and two eggs. A few days later there were four chicks, one of which was smaller than the others and died before the tower was set up. The nest was in a leafy Sandalwood tree at a height of six metres which was, although not unusual for inland areas, much lower than the usual fifteen to twenty metres height of nests elsewhere. It was a typical stick structure, lined with green leaves and in shade for at least part of the day.

The Black-shouldered Kite's reaction to human intrusion varies widely. Some will quietly fly off and watch from a distant perch, others attack fiercely - often emitting a harsh "kairr". The birds from the nest in the Sandalwood proved to be very aggressive, swooping to within a few centimetres as we climbed the tower or the nest-tree.

Most of our filming was done at a distance of three and a half metres from the nest. Feeding took place in the first three hours after sunrise and again from late afternoon until dark, indicating this species is to some degree crepuscular. Their diet consisted mostly of mice.

For the first two weeks the female brooded the chicks constantly - both day and night. Day brooding was essential because direct sunlight fell on the nest for several hours. After about two weeks the chicks were capable of moving into the shaded part of the nest.

This fully fledged chick had to be rescued from a Black Kite.

The female did no hunting at all for the first three weeks after hatching. Instead she called from the nest to the male, presumably for him to go and hunt up some food. If he didn't respond she would fly to where he was perched and chatter loudly for a few moments before flying back to the nest. He usually responded to this action and it was generally only a few minutes before he would be back at his perch with a mouse in his talons. She invariably saw him coming and would be at that perch only seconds after he alighted. She would snatch the mouse from beneath his talons and head directly back to the nest. We had

These thirty day old chicks are almost ready to leave the nest.

an uninterrupted view of the action and captured it on film.

The mice were fed in tiny pieces to the chicks for the first ten to fourteen days, after which they were capable of swallowing a mouse whole and often did so. The only time a mouse took more than a few seconds to disappear was when there was a tug-of-war between the chicks. This occasionally brought on a deadlock which would be broken by the intervention of the female, but she never ate the prey herself. Sometimes a chick would stand around with a mouse tail hanging from its bill, and another chick might try and pull the mouse out by the tail. A typical meal for the brood when a week or more old, would be three or four mice divided between them. Occasionally a small rabbit kitten provided the meal. The only time she ate anything on the nest was when the chicks were very small, when she would occasionally eat a piece she considered too large for the chicks to swallow. Her great care to ensure that only manageable size pieces were proferred was enthralling to watch.

The male came to the nest occasionally but he never brought any prey with him. He usually came when she was off the nest, but she would be back within moments of him alighting. It was felt that his presence at the nest was not to her liking as he always left the moment she returned. This behaviour is not unusual in some other hawk species.

Black feathers started to appear along the chick's wings when they were about a fortnight old and they were close to fully fledged at a month. We watched them leave the nest together at thirty four days. One was obviously not ready for flight as it only managed about a hundred metres before landing in the wheat stubble. A circling Black Kite made a dive for it, but we managed to rescue the fledgling in time and returned it to the nest-tree.

A Black-shouldered Kite alights on a stick above its nest.

The female Kite perches above her nest.

11

In November 1978 we worked a nest in a Mallee tree on a wheat farm. It was six metres above the ground and was so flimsy we bound it together with wire and secured it to the tree. The kites behaved like the others we worked. The male appeared to perform most of the hunting, then transferred the prey to the female which brought it to the nest and fed the chicks. He would perch on a dead stick above the nest and watch. The only prey we saw were mice.

It was impossible to detect any difference in size between the two adult birds although, as with other pairs we observed, the male's back was a slightly lighter shade of grey.

Although we haven't worked Black-shouldered kites since 1978, we often see them hunting over our properties, or along roadsides throughout Victoria and New South Wales, and are always captivated by these beautiful and graceful birds.

The nest and eggs of the Black-shouldered Kite.

The female Kite with her almost fledged brood.

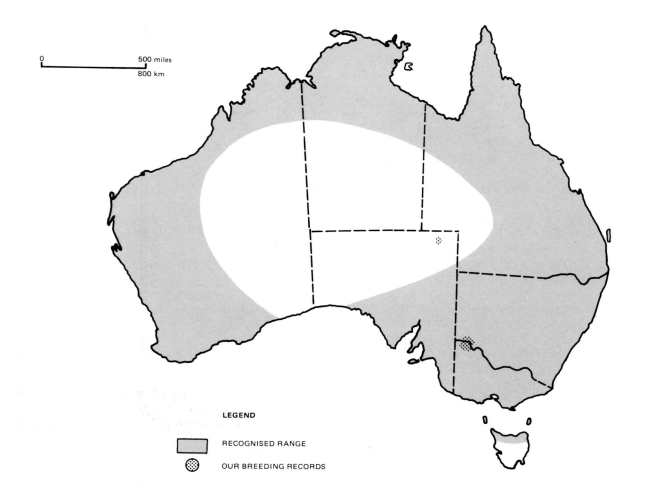

BLACK-SHOULDERED KITE *Elanus notatus*

elanos- kite (Gk); *notatum*- marked (L).

OTHER NAMES: Australian Black-shouldered Kite.

LENGTH: 330-380mm.

WINGSPAN: Approximately 900mm.

DISTRIBUTION: Common in open grassy areas in savanna and open woodlands, and around cultivated areas in most coastal and adjacent regions, where many birds appear to be sedentary. Less common inland, usually absent from the arid interior (except in high-rainfall years). Inland populations are migratory or nomadic. Not found outside Australia.

VOICE: Whistling **'chip-chip-chip'**; also a wheezing **'kair-kair'**.

PREY: We have recorded mice, small lizards and skinks, large insects, a frog, and a three-quarter grown rabbit. This kite is reported to take small birds.

NEST: Usually a small stick structure of about 300-350mm in diameter, though sometimes larger - particularly if an existing nest is added to. The centre cup is 150-200mm in diameter and lined with green leaves. A fork in a vertical tree branch is the most favoured site, usually at a height of 15-20 metres, though generally lower in drier regions. In the Mallee of north-western Victoria we found several nests 5-7 metres above ground.

EGGS: Usually three or four, 40 x 30mm. They are round ovals, slightly glossy, whitish with large blotches and smears of dull brown, chocolate and red-brown. Of fifteen clutches recorded by us ten consisted of four eggs and five of three. Egg laying occurs over a long period, from May to November. Early-breeding birds often nest twice. We recorded egg laying from August to mid-November. The incubation period appears to be about 30 days, and chicks fledge in 34 days.

13

LETTER-WINGED KITE
Elanus scriptus

When perched, the Letter-wing appears to be almost identical to the Black-shouldered Kite. However, if studied closely, a number of differences can be detected. The legs of the Letter-wing are pale flesh coloured or very faintly yellowish, while those of the Black-shouldered Kite are bright yellow or yellow-orange. The grey cere of the Letter-wing differs from that of the Black-shouldered Kite, which is yellow. More subtle differences occur in the plumage of the two species - that of the Letter-wing being softer and more owl-like.

The eye and black eye patch are noticeably larger in the Letter-wing, giving it quite an owl-like appearance at times. At least one observer has disputed this as an identification factor. We would concede only that the proximity of the subject and the angle from which it is viewed may emphasise the owlishness or otherwise.

In flight, the Letter-wing is unmistakable, having a line of black feathers running the length of the underside of the wings. This often forms the letter W or M - thus the English and Latin names. Wing action is also distinctive, being somewhat like that of the Caspian Tern, the wing action being deep and loose, and the flight buoyant.

We have found this bird to be more nocturnal than diurnal. Some have described it as crepuscular, (active at twilight), but we would insist that it's nocturnal in its feeding habits. We have never seen the chicks being fed during daylight hours, nor have we seen the chicks show any interest in the return of the adults to the nest during the day. Only toward evening do the chicks show any animation. We worked this species extensively from 1974 to '76 and most of the work was done at night and often on moonless ones.

David Hollands, along with a party of bird watchers had flown to Birdsville in August, 1974. He had photographed the species to the north of that town and on his return had contacted us. We took off for Birdsville within hours of receiving the

The soft plumage and large eyes give the Letter-wing an owl-like appearance.

information. It was actually a few minutes after midnight when we pulled away from home. Daughter and sister Pamela, also afflicted with an inherent urge to roam insisted on a berth. She was designated chief cook.

We went via Renmark, Morgan, Burra, Hawker, Leigh Creek to Marree. The road was rough in places, much of it unsealed, but we were in Marree by noon. Enquiries there afforded the information that the Birdsville Track was open to four-wheel-drive vehicles. With only a conventional drive Ford utility with our tower atop, we were naturally a little apprehensive about our chances of getting through, but having plenty of food on board, and sufficient petrol to take us fifteen hundred kilometres under normal driving conditions, we were prepared to take calculated risks to get some shots of such a unique species.

Over the years the Birdsville Track has grown into something of a legend in Australian history. It stretches for almost five hundred kilometres from Marree, just north of the Flinders Ranges, to Birdsville, a few kilometres north of the Queensland border. It is an area of violent extremes, perhaps best described by Henry Lawson's words from ''The Never Never Land'':

'A blazing desert in the drought,
A lakeland after rain'.

The Letter-winged Kite in flight.

The first exploration of the area was made by Charles Sturt's 1844-45 expedition. The men suffered terribly from the lack of water and extreme heat. Temperatures of 157° F. (69° C.) in the sun, and 132° F. (55° C.) in the shade were recorded. Sturt wrote: "The sky . . . was without a speck, and the dazzling brightness of the moon was one of the most distressing things we had to endure". [1]

Burke and Wills passed through this area in 1860 on their ill-fated journey north from Innaminka to the Gulf. At first they didn't encounter conditions as harsh as those experienced by Sturt. Wills wrote in his diary. "I was rather disappointed to find the desert nothing more or less than stony rises . . . and many a sheep run is, in fact, worse grazing than that". [2] However Burke and Wills eventually perished near Innaminka.

When we first arrived in the area, record floods had filled Lake Eyre for the first time in seventy years and the Track was extensively damaged. In some places creeks flowed over it, and we thought it prudent to check the depth of the water and the road surface beneath it by first walking through wherever we were in doubt.

It is a place that beguiles the unwary traveller. At Marree a signboard warns motorists of the dangers of driving on the Track, and requests them to report journeys to the local police station before continuing. Even so, many have perished. The Track exacts its retribution on those who are careless of it, those who do not respect it. Perishing of thirst was not likely to be one of our worries at this particular time.

Progress was naturally slow, but we reached Etadunna homestead just before dark. Here a large iron cross stands as a memorial to the Lutheran mission established on Cooper's Creek sixteen kilometres to the north, over a century ago. It had actually started in 1867 and had hung on through years of drought and the harshest conditions imaginable, before closing finally in 1917.

At this point we detoured, as the Kopperamanna crossing was flooded. After what seemed an interminable age, crawling along a very rough winding track, we reached Cooper's Creek just upstream from Lake Killamperpunna. Here the stream was relatively narrow, and deep enough to run a barge.

Crossing the Cooper by barge.

Cooper's Creek is better known as The Cooper. Through its tributaries it drains a vast area of north-central Queensland. Even so, its waters seldom reach Lake Eyre as it has many lakes to fill along its course. It is unique as a 'creek' in having as tributaries, two rivers, the Barcoo and Thompson.

The Cooper was about two hundred metres wide at this point and a notice board stated that operation of the barge was permitted only during daylight. The operator at that time interpreted daylight as eight till five. "You can camp out there", he said, with the wave of a hand in a southerly direction. We drove off and a few moments later the headlights picked up a galvanised iron shed. Being by then a bitterly cold night with a lazy wind - it preferred to go through, rather than around one - we threw our swags down in the lee of that building.

Next morning we had the billy on the gas-ring set on the tailboard of the utility when the barge operator strode over. In tones as chilly as the morning he said, "I hope you blokes don't smoke near that bloody shed. It's full of bottled gas!" We assured him of our due respect for property, the fact that none of us smoked and our profound and healthy respect of our future - we had no urge to be blown up.

The tower, bolted to its frame on top of the utility, stirred his curiosity and he gruffly enquired as to its function. Our conversation turned to birds and he asked if we knew Norman Favaloro, a well-known ornithologist, in Mildura. We replied that we came from the same area and we were hoping to band some Letter-winged Kites for him. In an attempt to thaw the ice a little more, we asked if he knew any of the Cuppers around Mildura. He said he once knew a Jack Cupper who served in the same battalion as himself in the Middle East.

Thirty three years earlier the two had last seen each other in the Libyan Desert. Ian Farnell held out his hand and Sturt's Stony Desert seemed a lot warmer as we shook hands.

The barge was powered by two out-board motors, mounted one each side, and was used to transport cattle as well as vehicles up to the size and tonnage of a prime-mover for a stock transport. There could be long delays, especially on cattle shipping days. Ian put us across at 07:30, half an hour ahead of his usual schedule. The thaw was complete!

We left the Cooper and drove through the Natterannie Sandhills and back to the main track, now interspersed with boggy patches which we carefully avoided. Occasionally we saw what appeared to be smoke rising ahead of us. It turned out to be steam rising from artesian boreheads. The artesian basin is tapped at about a thousand metres and the water is near enough to boiling point. In later, drier trips we always filled our containers where water flowed across the Track from the Mungerannie Bore. It was excellent drinking water.

15

On Sturt's Stony Desert, the gibber stretches to the horizon.

The gibber stretched away to the horizon. This was the desert Sturt painfully explored, on which only ants and lizards moved. Sturt's Stony Desert is a relic of a former seabed. After the water receded the hard sedimentary layer that had formed while under the sea, shattered and weathered to its present form - a vast hot plain cobbled with red-brown stones. The Aboriginals called these stones gibber. In the frosty nights, dwarfed by the vastness of the sparkling Milky Way, Henry Lawson alone could describe that plain:

And camped at night where the plains lie wide
Like some old ocean's bed,
  The stockmen in the starlight ride
Round fifteen hundred head.

More appropriate for this particular time perhaps would have been the line:

'To the skyline sweeps the waving grass', for in the years from 1973-75 even the desert bloomed. In many places the grass grew in the cracks between the stones, and the creek beds that traverse the area at varying intervals were waist to more than head-high in grass and herbage.

This unprecedented flush of plant life coincided with, or perhaps triggered, an irruption of Long-haired native rats. So vast was this horde that many of the creekbeds were honey-combed with their burrows which made walking extremely heavy going, as one broke through to them with almost every step.

Explorers and settlers commented on some of their past irruptions. In 1860 Burke and Wills suspended their equipment from trees to keep it safe from them. During the irruption in northern Queensland in 1869-70 they attacked almost anything capable of being chewed, even gnawing the greenhide hobbles off the horses at night. In 1887 there was an enormous migration of them near Lake Eyre and John Bagot reported their numbers went into the millions. [3]

The rats feed on seed and herbage and were popular in the diet of the Aboriginals. Jim Hogg, who accompanied us on later trips, found many of their work-sites and artifacts. Jim's long grey hair and beard, together with his aquiline features, must have had tourists along the Birdsville Track thinking he was an old Afghan hawker tracking his camels. He wasn't much value in searching for raptors - his eyes usually roved the ground - but he never failed to amaze us with his discoveries of the Aboriginals past occupation of almost every area in which we worked. In many of the middens uncovered by the huge floods in 1973 he found quantities of mussel shells suggesting the area once sustained more surface water of a permanent nature.

In turn the rats were food for birds of prey. In addressing the Royal Society of New South Wales on 1 May 1918 Dr. Cleland lectured at length on irruptions of the native rats and the kite's usefulness in their control.

About two hundred kilometres north of the Cooper, at Clifton Hills Station homestead, we took the Outside Track as the main one to the north was under water. One hundred and forty kilometres south of Birdsville a bird flew out of some Coolabahs lining a creek bed by the side of the road. It was a Letter-winged Kite, the first we'd seen. We were about to learn something of their habits, for as we pulled up to get a better look at that lone bird, a whole colony of them took to the wing from the Coolabahs. We learned later that we had passed several colonies along the way. They seldon fly during the day unless disturbed. At this colony nearly every tree had a nest, if not a Letter-wing's then some other raptor's.

A Long-haired Native Rat - a major food source for the Letter-winged Kites.

The site of a Letter-wing colony along the Birdsville Track.

We found Black Kites, Brown Falcons, Spotted Harriers, Australian Kestrels, and later Black Falcons. The 'Birdsville Track' could not have been more aptly named. To us it was paradise, a veritable Garden of Eden, and we were to return to it again and again in the weeks and years to come. So often did we do the two thousand seven hundred kilometre round trip that almost every rut or hazard was memorised and thus avoided.

We did however go close to a major breakdown one night when we unknowingly knocked a hole in the sump of our vehicle when we struck a pile of stone in the middle of the track, just before we reached the Cooper. As we pulled up at the Creek the motor appeared to start missing. The hood was raised to see if a plug wire was off and a small pool of oil was spotted on the ground beneath the motor as it gleamed in the moonlight. The loss of oil had caused the hydraulic tappets to drop their oil and rattle, thus giving what we thought was an engine miss. By covering the hole with a piece of opened out plastic hose, kept in place with self-tapping screws, we were able to continue our journey. Ian Farnell supplied the tools and oil for the job.

We returned to the area in May 1975 and made camp close to the Track, on Damperannie Creek. We were busy preparing an early evening meal when we glanced up, and were startled to see a string of camels within a few metres of us. So silently had they moved, even over the gibber, we had heard no sound of their approach.

The camels carried a small group of tourists, led by a couple of enterprising cameleers, and had travelled across the southern edge of the Simpson Desert to Birdsville and were now on their way south to Marree, travelling at an average rate of around thirty kilometres per day. They camped for the night on the other side of the Track from us. The rats didn't worry them; strangely enough, we found they seldom did during the first night at a new camp, but got progressively more troublesome on subsequent nights.

A female Kite alights beside her nest during daylight.

We decided to shoot some cine film on the Letter-wings feeding their chicks at night. We were working a nest in a Beefwood, photographing with still cameras at close quarters. The adults were behaving quite well, so we sited a petrol engined generator about a hundred metres downwind of the nest. A plastic-covered cable carried the power to the lights mounted on the front of the hide. We worked quite close to the nest, about two and a half metres, to ensure there was enough light.

The generator was started at dusk and the hide entered. It was 21:00, and pitch dark, before the first bird was heard to alight. A hand torch was used to confirm the presence of the female in the nest with a large rat, and she was about to feed her hungry brood. Keeping fingers metaphorically crossed, the lights were switched on. Apart from a slight hesitancy on the part of the female all was well. She continued to tear apart a very tough rat and fed the chicks large pieces with skin and fur attached. The size of the pieces had to be seen to be believed, almost a quarter of the rat being gulped at a time.

Striking visions of a by-gone era, a tourist camel train makes its way down the Birdsville Track.

After dark a brood of five chicks awaits the arrival of the adults with food.

Filming was suspended about midnight with the intention of catching up on some sleep and doing a bit more filming before dawn. After a few hours trying to sleep while rats held races along our sleeping bags, sampled our hair, and anything left lying around, we arose and returned to the nest. The generator was started but the power was not getting through to the lights, and it was found that the rats had chewed through the power cable.

They are apparently quite partial to the plastic covering, but we were intrigued with their ability to gnaw through the wires as well. By the time we'd found and fixed the break in the line it was breaking day and the prospects of getting any more feeding sequences were remote as the rats stay in their burrows and the Letter-wings also rest up.

Night filming Letter-winged Kites with the aid of powerful movie lights.

This Letter-wing chick's first flight ended in a low bush below the nest tree.

The breeding and feeding habits of the kite appear to have evolved through close association with the native rat. We believe their very existence is dependent to a large extent on rat irruptions. When the rat population increased in 1974 it was paralleled by an increase of Letter-wings as they bred continuously. In June 1975 we found flying juveniles as well as eggs and chicks at all stages of development. In December we found eggs ready to be hatched. The most likely explanation of this rapid increase in the colonies is the birds breed when they are relatively young. Some of the breeding birds we observed still showed traces of juvenile plumage. Extensive banding and close observation is necessary to confirm our conclusion.

By June 1976 the rat irruption stopped as suddenly as it began. The flocks of kites and other raptors had decimated the rat hordes. The Letter-wings stopped laying and among those last to breed we noted the clutch size was smaller. Prey became scarce and it was obvious the birds had to move or die of starvation. As the last of the chicks took to the wing the Letter-wing colonies moved out *en masse*. Some other species left it too late. We found many Black Kites too weak to fly and they were usually light in weight. The raptors in general scattered throughout Australia - the Letter-wings heading south, and many were observed in Victoria.

The first few appeared near our home on 2 December 1976 and within a few days a colony of twenty-eight birds was established. Some of them showed traces of their juvenile plumage. They stayed for almost four months - possibly feeding off locusts and mice, which were close to plague proportions, small rabbits and rabbit kittens. However, we had no proof of their diet because they did not breed there, nor at the sewerage farm at Werribee in southern Victoria where many were sighted. Although mice were plentiful they did not appear to thrive, there were numerous reports of dead and dying birds. [4] They had moved to an environment alien to their natural habitat, the hot and dry regions of central Australia. They seldom breed anywhere else.

Writing in the **South Australian Ornithologist** in 1934 J. Neil McGilp stated that the Letter-winged Kite does not come further south than a line drawn latitudinally through Farina, South Australia (30° 02' S.). [5] He has been proved substantially correct since they do so only after such large-scale irruptions as the one described here. A. C. Cameron's 1974 Sunbird paper [6] outlined observations of the species nesting in Coolabahs in western Queensland between latitudes 23° S and 26° S and we observed colonies near Dulkaninna, South Australia (29° 02' S. 138° 27' E.). Like Cameron we found many nests in Coolabahs, but where there was a choice between Beefwood and

Coolabah, the Letter-wings invariably chose the Beefwood as their favourite nesting tree. Most of the nests were in colonies with seldom more than one pair of birds to a tree. Where there was a newly constructed nest in the same tree as one containing advanced chicks we're certain it belonged to the same pair preparing for their next brood. Occasionally we found a pair breeding in isolation many kilometres from a colony.

In 1978 we noticed a single bird, possibly remaining from the colony established near our home two years earlier. It flew low overhead as we were setting up at a Black-shouldered Kite nest and for a moment we wondered if this was a case of interbreeding species. The solitary bird, a member of an inherently gregarious species, joined the pair of Black-shouldered Kites which are, after all, its closest avian relatives.

Letter-winged Kites were well-publicized between 1974 and 1979, but much of the later literature concerned their activities in Victoria. Most of the available evidence indicates the majority of the kites projected their exodus to the south. Why south? Do any return to the inland? Do some breed somewhere each year to perpetuate the species between rat irruptions? Or does the survival of the species rely on a nuclei of longeval birds? Although we learned much about them over the years many questions remain. Given the size and inaccessibility of their habitat they may take some time to fully answer.

And yet some of our observations may have unravelled part of the mystery. In the late summer of 1979 we found a colony of seven pairs of Letter-wings breeding in an area known as the Cobblers* a few kilometres south of Montecollina Bore on the Strzelecki Track. Since rats do not come south of the Cooper we were surprised to find them. It was the first we had recorded breeding since 1976 and indicated that perhaps small colonies of them breed intermittently between the rat irruptions. The nests were from three to eight metres above ground in Coolabahs, the only trees in the area, and were lightly lined with eucalyptus leaves. The clutches ranged from two to four eggs.

A typical nest and eggs of the Letter-winged Kite.

We reconnoitred along the Birdsville Track as far as Damperanie Creek on 'Clifton Hills' Station without sighting another Letter-wing - a great contrast to the profusion of Letter-wing colonies found along there on our first visit five years earlier. Many other species were breeding however, and some chicks were close to fully fledged. We estimated that breeding would have begun in February, no doubt triggered by the heavy rains in January when almost 300mm were recorded in the Innamincka-Birdsville area - twice the annual average in one month. It is only in the Inland that the breeding season fluctuates so widely. Prolonged adverse conditions, characteristic for the area, have made many species opportunity breeders.

On 14 May 1979 we returned to one of the nests we had inspected which had contained four eggs. We set up the tower and found two chicks about a fortnight old. The generator was dug into the far side of a knoll about a hundred metres away to keep noise at the nest to a minimum. It was started up just after sunset. The female alighted close to the nest about half an hour later and called to the male perched in a dry tree nearby, silhouetted against the after-glow. He flew off and returned in a few minutes to a noisy meeting with the female

Along the Birdsville Track, the tower at a nest in a Beefwood.

* *An area of predominantly white sand which has been swept into countless small hillocks by wind action. From a distance it sometimes resembles a flock of sheep. It probably derived its name from shearers who used to refer to the roughest and toughest sheep in the pen as a 'cobbler'. It was invariably the last in the pen to be shorn.*

A view of the Cobblers nest site, with the generator and tower ready for night filming.

near the nest, then she flew in carrying a large mouse. The lights were switched on and some footage was taken of her tearing the mouse to pieces to feed the chicks. She appeared uninhibited by the lights, although we learned to switch them on after the birds alighted and not before or else they would be blinded and crash into the foliage.

A fortnight later we returned and found the brood reduced to one. We checked the other nests in the colony and were baffled to find that of the original seven nests containing at least sixteen eggs, only four hatched and two chicks survived to become fully fledged. One lone chick was able to swallow mice whole, though sometimes only after a herculean effort.

The weather wasn't always favourable to us. One night we decided to make efficient use of the lights and set up two towers at the nest. We were filming when it began to rain. Trickles of water found their way into the switches. One or two shocks later we were out of the hides and descending the ladders in record time - one moment in total darkness, the next in dazzling light as the water entering the switches activated the lights. The birds, meanwhile, sat stoically, and silently endured our antics.

The female Letter-wing with her almost fledged chick.

21

A Letter-winged Kite alights above its nest.

After the female finished day-brooding she perched either on the edge of the nest or on the dry branch immediately above it. We would film her as she alighted on her chosen perch. On one occasion, while waiting for her, some Little Corellas flew in and perched on the branch, screeching and performing so beautifully that some shots of them had to be taken. Some others alighted on the canvas top-cover of the hide. Their screeching, mere centimetres away, wasn't appreciated at all. A slap on the canvas beneath one of them sent them screeching on their way.

A pair of beautiful, but mischievous Corellas land above the Kites' nest.

At our next visit we found the two surviving juveniles on the wing, but they were still being fed on their nests at night. The parents of the one we'd been working built a new nest close by before their chick had flown from the other. It was lined with teased-out pellets probably disgorged by the birds. Apparently the nests in the new colony are lined with eucalypt leaves for the first round of breeding. Thereafter teased-out pellets are used for subsequent broods. It is a logical way of lining the nest because pellets are often disgorged into the nest by the brooding bird and later by the chicks. They are almost entirely comprised of fur, with a few slight traces of bone.

Pellets from this and other nests in this colony were sent to David Baker-Gabb at Monash University for microscopic examination. The fur proved to be predominantly that of the House Mouse, with traces of fur from the Fat-tailed Dunnart also present in some pellets.

The Kites continued to occupy the five nests which had failed to produce a brood. There were frequent flights to mate either in their respective nest-trees or in trees nearby. These flights were usually made by both birds climbing to relatively high altitudes, their wing dihedral angles forming tight 'V's above their bodies and their wing-tips fluttering giving them forward and upward motion. They chattered continuously as they performed this manoeuvre. Initially we were convinced the birds were reacting to our presence, but when we observed them flying off like this spontaneously, after the hide had been occupied for hours, or when we were well away from the colony in our camp, we felt sure they were quite often display flights, although there was little to distinguish them from disturbance flights, except that copulation took place often after the flight.

About twelve kilometres away we found a second colony of Letter-wings with about the same

Female Letter-winged Kite.

number of birds and nests. The approach, through sand drifts and salt-encrusted seepage patches, was hazardous enough for us to prefer working at our present site - despite the fact that the tower was visible from the track and attracted curious travellers. The birds were apparently preparing to breed, but there was no other evidence in the second colony of earlier attempts as there had been with the first.

During our search for the Grey Falcon we passed through the area on 16 July 1979 and paused to inspect the nests for breeding progress. There were nine nests with clutches ranging from two to four eggs making a total of thirty-one. We set up the tower at a nest which two months earlier held three eggs which failed to hatch. There was now another clutch of three eggs. Anticipating some damage from the Corellas we used eight guyropes on the tower instead of the usual four.

Returning from a search along the creeks on the Birdsville Track about a week later we discovered the tower down, but fortunately undamaged. The Corellas had nibbled through the four ropes on the windward side and the falling hide narrowly missed taking the nest with it. We could well imagine that colourful and raucous scene as the

tower toppled, with the mischievous Corellas screeching triumphantly before beating a hasty retreat. After that we used steel guy-ropes.

On 8 August we were back, and set up at a nest containing one chick and two eggs. Our attempts to film the chick being fed were frustrated by both adults fleeing every time the lights were switched on. After a few of these unsuccessful attempts we abandoned the hide. If we had persisted we may have caused desertion.

We returned nineteen days later. So had the Corellas. The canvas covering on the hide was in tatters. With an electric drill, a pop rivetter and some spare towels we repaired the hide sufficiently to meet the demands for the coming night's work. No sooner had we finished than the Corellas were back. We found it was easier to occupy the hide and keep them on the hop by pushing a pin through the canvas into their feet. Fortunately the antics of the Corellas didn't disturb the Letterwings, although they no doubt puzzled them.

There were now three chicks in the nest, close to three weeks old - a one hundred per cent hatching and survival rate. Of the nine nests containing thirty-one eggs nineteen chicks survived to advanced fledging. Whatever the cause of the poor hatching rate in the first round its effect was diminishing. Our earlier bafflement gave way to some speculative thinking. The high infertility rate may have been due to the number of non-breeding years, caused by such adverse circumstances as a shortage of suitable prey. The high incidence of egg infertility pointed perhaps to a diet deficiency affecting the fertility of the birds, or possibly a long period (perhaps several years) of sexual inactivity had caused the reproduction organs to atrophy. With stimulated activity the rate of fertilisation improved rapidly, hence the much improved breeding success with the second round of breeding. Even so the results of the second round were considerably below anything we saw in 1974-75.

Over the following few nights we obtained some excellent footage of the chicks being fed. The adults settled down and behaved naturally, regularly bringing in mice which the chicks, after some lively tussels, would swallow whole. These birds have an uncanny ability to locate mice, even in grass, on the darkest of nights. One mouse brought in was still shrouded in the grass which had been snatched along with it.

On several occasions we noticed the eggs hatched progressively. On 8 August we inspected a nest containing one fresh egg. A second inspection on 13 September revealed a freshly hatched chick and three eggs - indicating an incubation of thirty-six days, longer than anticipated. Further study is warranted, however, on this hypothesis.

Male Letter-winged Kite.

It is quite possible that incubation did not start with the laying of the first eggs.

In **Sunbird** in 1974 A. C. Cameron noted a high degree of inter-species tolerance when nesting. We found it quite common - especially where such prey as native rats, mice and rabbits are plentiful and trees are not. This was the case with the Letter-wing colony. In one Coolabah, apart from the Letter-wings, there were three other active nests; an Australian Kestrel less than a metre away, a Black Kite two metres and a Raven perhaps three.

When we visited the Cobblers in April 1980 the land had changed. The hot, dry summer had taken its toll of the vegetation and animal life. Six months earlier a luxuriant oasis of green Coolabahs sustained at least thirty avian species. Now the trees offered merely broken shade, there was little else to relieve the surrounding desolation:

To the skyline sweeps the waving grass,
Or whirls the scorching sand -

A phantom land, a mystic realm!
The Never-Never Land. *(Lawson)*

The pair of Letter-wings react to the close approach of a Black Kite.

An almost fledged Letter-winged Kite in an aggressive pose.

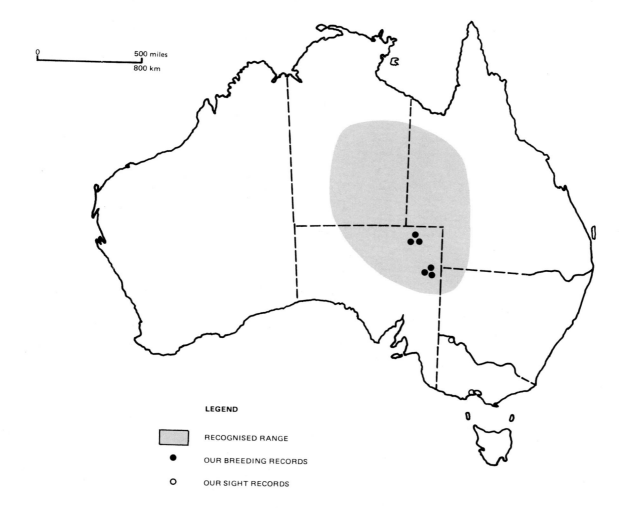

LETTER—WINGED KITE *Elanus scriptus*

elanos- kite (Gk); *scriptus*- written (L)

OTHER NAMES: White-breasted goshawk.

LENGTH: 350-380mm. Female very slightly larger than male.

WINGSPAN: Approximately 900mm.

DISTRIBUTION: Generally rare, nomadic; usually scattered throughout the dry, sparsely timbered plains of the interior. Irruptive, becoming locally very numerous during irruptions of native rats in the interior. Following irruptions, may spread over vast areas, sometimes reaching coastal regions. Not found outside Australia.

VOICE: The alarm call is a high-whistling 'chip-chip'. At other times a rasping, 'karr-karr', or a slow harsh chatter.

PREY: When native rats irrupt they constitute the kite's total diet. At other times mice, large insects and perhaps small reptiles. We have recorded rats and mice only.

NEST: New nests are similar to those of the Black-shouldered Kite, though only sparsely lined with green leaves. These leaves are soon overlaid with a thick layer of fur, usually from teased-out pellets. Nests are often, though not always, placed in the top of a tree. At times they may be built on top of an existing nest; we've seen a Letter-winged Kite building on a Black Kite's nest. Nests observed by us ranged in height from three to eight metres.

EGGS: Clutch size is dependent on existing conditions and ranges from two to six eggs, though four or five is usual in times of good food supply. They are rounded ovals, 44 x 32mm, slightly glossy, bluish-white, heavily blotched and smeared with chocolate and red-brown. During the native rat irruption of 1974 to 1976, when egg-laying was continuous, nearly all clutches we saw were of four eggs, with two clutches of five. In 1979, of 24 clutches which subsequently proved to be completed, ten contained four eggs, thirteen contained three and one only two eggs. These were laid from April to October. Because of the sometimes erratic timing of egg-laying, and apparent disappearance of some eggs, incubation periods were hard to establish. In one case 36 days was indicated, while in another it appeared to be less than 25 days. Fledging period is 30-35 days.

BRAHMINY KITE
Haliastur indus

This handsome fishing eagle in snow white and rich chestnut plumage is tolerably common along the coastal regions of tropical Australia . . . . [7]

As was written around the beginning of this century, the Brahminy kite inhabits the coastal regions from north-west Western Australia to the northern coast of New South Wales. We've also found it some distance inland along rivers in Queensland. In Maryborough, Queensland, we've seen it swoop down to pick up scraps from the nature strip beside a busy street in the heart of that city. It is fairly common along the Queensland coast where it may be seen sweeping the creek and river estuaries and mud flats at low tide, in search of stranded fish and other marine animals, dead or alive.

Brahminy Kite in flight.

Around Australia, the kite's close association with coastal regions has led to it often being called the Red-backed or White-headed Sea-eagle. Elsewhere throughout its range, particularly in rice-growing areas of Southern India where it is reportedly the commonest of all raptors, its scavenging habits are more evident.

We first photographed this raptor in July 1974. We found a nest near Hay Point, south of Mackay, Queensland, in a large eucapypt, eighteen metres above ground. The female was incubating eggs, and both birds called plaintively while the tower was erected, although she returned to the nest within minutes of the hide being entered. During the following few days they proved relatively tame. Once on the nest she was hard to flush, and sat tight as we entered or left the hide.

The nest contained two eggs, but only one chick was eventually reared. This was the case with the three other nests we worked. We never found a rotten egg in a nest along with a chick and were unable to ascertain whether, in any of the nests, both eggs hatch.

In August we set up at our second nest for the year, it was on the lower reaches of the Mary River in south-eastern Queensland. It was at a height of ten metres in a Stringy-bark. The female was on two eggs and, like the one we'd worked the previous month, was back on the eggs soon after the hide was entered. We took a few photographs and left.

A clutch of Brahminy Kite eggs.

We returned on 4 October and found one chick about three weeks old, indicating the eggs were laid early in August. The female brought a fish to the nest and immediately flew off leaving the chick to feed itself, apparently for the first time, going by the manner in which it was managing the task. The proximity of the hide probably deterred the female from remaining to feed the chick.

At this nest, as with the previous one, we were surprised that the female was reluctant to remain at the nest with her chick, whereas, when incubating eggs, she was completely undeterred by our presence. Most species are more attached to the chicks than to the eggs, and it was this fact which encouraged us to avoid a nest until the eggs hatched. Some species will return to their eggs and continue incubating while an observer is

A half grown Brahminy chick waits for the female Kite to begin feeding.

29

Using the tower as a ladder to enable some ''gardening'' to be done prior to beginning photography.

standing near the nest-tree. Not all will, however, and as we learned more about each species we grew to recognize the danger signs and adjust our approach accordingly. Later we worked many species when they were incubating and recorded valuable information on their behaviour.

The chick made little progress feeding itself and after half an hour the female returned and fed it. Fish was the only prey brought to this nest. They appeared to be dead, probably washed ashore. After the chick's appetite was sated, the female carried the remains to a nearby tree and, using a large horizontal branch as a feeding platform, fed herself.

We worked another nest near Hay Point in 1976. It was in the outer foliage of a eucalypt so we set up the tower between the centre branches. We had a clear view of the nest without having to remove any of the surrounding foliage and, since the hide was only four metres away, we watched for any reaction from the adults. The chick was no more than a fortnight old, but the female would not come to the nest while the hide was occupied. We lowered the tower and she came in and fed the chick. We waited for two days then slowly raised it again. While the hide was vacant it was accepted, when it was occupied she remained wary and would not come to the nest. All kinds of stratagems were devised to induce her return: a 'seeing-in' party was used constantly, and the hide was draped inside to blur any shadows or silhouettes. All to no avail. We wondered whether she could count and therefore notice that the number of people climbing the tower to the hide exceeded the number leaving it. Then, one day near the end of a stint in the hide, there was a sharp shower of rain. Within moments she flew in. She was probably watching the hide more than the nest, because she landed awkwardly, practically falling into the nest, fortunately leaving the chick unharmed. She brooded until the rain stopped, then left the nest, but from then on she appeared to lose all fear of the hide, and behaved naturally.

Strong winds blew through the area daily, and since the nest rested on a relatively thin branch, it swayed in an arc of at least a metre. We marvelled at that bough's strength.

In July and August 1977 we observed this same pair incubating eggs in a second nest, about eighty metres away from the first. They'd unfortunately chosen a tree in an area earmarked for development of a recreational project. For two days, surveyors and their teams tramped through the bush around the tree, and the female rarely returned to brood. The eggs did not hatch.

This saddened us. Obviously the Brahminy Kites, as well as a pair of Ospreys nesting nearby, would have to move. We could also appreciate that much of the natural charm of the area, one of the most beautiful locations along the Queensland coast, would soon be destroyed.

The female Brahminy Kite with her small chick.

The picturesque mouth of Louisa Creek, where we camped when working the Brahminy Kites.

Our fears seemed confirmed when we saw no sign of the birds in the area in 1978 and again in 1979. We looked for them on our trek north in 1980 in search of the Red Goshawk, and were thrilled to discover they had returned. We set up at the nest, and found the birds easy to work. Our expectations about future work at this nest were high, and we planned to return. The events surrounding that other search, however, brought our study of this pair to a standstill.

Our difficulty in getting the birds to lose their fear of us and behave naturally could be overcome with greater patience and perseverance. The rewards of this approach were strikingly demonstrated to us by Mrs. Judith Little of Shute Harbour, Queensland, who has perfected a call which brings the Brahminy Kites to her home. Since we were strangers to them, Judy asked us to watch from the shadow of her house. She laid out three or four fresh loin chops on a small table, and gave her call - a perfect rendition of the Brahminy's plaintive **'pee-ah-h-h'**. Apparently it carries as far as the bush 'Coo-ee,' for in less than a quarter of an hour birds appeared, coming in from across the harbour and over the nearby ranges, to circle high above the table. Then, one by one, the kites peeled off and dived like Stuka bombers to swoop, snatch a chop and fly up the mountainside to perch in a tree and eat it. As the chops disappeared Judy replaced them. The birds weren't inhibited by her standing a few metres away from the table. It was a fascinating exhibition of what can be achieved with the correct approach and the wherewithal to obtain loin chops.

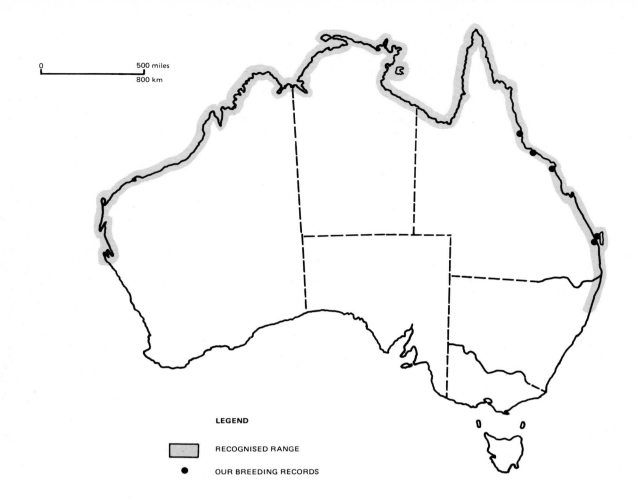

0   500 miles
800 km

LEGEND

RECOGNISED RANGE

●   OUR BREEDING RECORDS

BRAHMINY KITE   *Haliastur indus*

hals- sea (Gk); *astur-* goshawk (L); *indus-* India.

OTHER NAMES: Red-backed sea-eagle; White-headed sea-eagle.

LENGTH: 430-510mm. Female slightly larger than male.

WINGSPAN: Approximately 1200mm.

DISTRIBUTION: Common and sedentary near beaches, mudflats and mangroves of coastal north-west, north and north-east Australia and off-shore islands; extending inland along larger rivers. The Australian subspecies *H. i. girrenera* also ranges through New Guinea and neighbouring islands, while other subspecies extend through Indonesia, South China, tropical Asia to Sri Lanka and India.

VOICE: Plaintive, quavering **'pee-ah-h-h'**.

PREY: Fish, frogs, crustacea, reptiles and insects. Carrion and offal is also eaten. We have seen only fish brought to nests we have studied.

NEST: A rough structure of sticks 400-600mm in diameter, with centre cup lined with seaweed, rag, paper, leaves, grass or just finer sticks, placed in the fork of a tree 10-20 metres above ground. We have found most nests were newly built each year; the exceptions being substantially larger structures. In North Queensland we found nests in transmission towers.

EGGS: One to three form the clutch, usually two, 51 x 40mm. They are rounded ovals, dull white or bluish-white, finely streaked and freckled with purplish-brown. Of the six clutches we have recorded four consisted of two eggs and two were single eggs. A seventh nest contained a half-fledged chick. Eggs are laid from April to September in northern and north-western Australia, and from July to October along the east coast. Our records, from east-coast nests, indicated egg-laying from July to September.

SQUARE-TAILED KITE
Lophoictinia isura

The Square-tailed Kite is generally recognized as being thinly scattered throughout Australia in open forests and woodlands. It is not found in heavily forested or treeless areas. Considered to be a bird of the Inland, we found it only in wooded coastal areas and the sub-interior.

There were few recorded sightings over the years. Gould found it breeding at Scone, New South Wales, in 1839 and his contemporary, Gilbert, found it in Western Australia.[8] Beginning late in the 19th Century egg collectors were particularly active, and Square-tailed Kites' eggs were highly prized. Many eggs, especially in Queensland and New South Wales, were taken.

Little was recorded thereafter until 1950, when A. C. (Cec) Cameron began systematic records of the species breeding in the vicinity of his property 'Rockwood', near Chinchilla, Queensland. From 1969 he found them along the Wieambillia Creek, within a few hundred metres of their original nest. They sometimes used the same nest a second time.[9]

In 1974 we contacted Mr. H. T. (Herb) Condon, Curator of Birds at the South Australian Museum, Adelaide, who directed us to a site near Port Lincoln where the species had been found nesting in the past. We searched the area thoroughly, but found only a single bird. We were shown a nest,

on a substantial horizontal bough, where a pair had nested two years earlier.

In July 1974 we were travelling north on the Bruce Highway near Bororen, Queensland, when we sighted a pair of Square-tails circling over a forest. After a short search we found them nest-

The Square-tailed Kite in flight.

building at a height of twenty-two metres in a Lemon-scented Gum. They carried sticks and placed them on the nest, then one worked them into position by sitting and wriggling as if making itself comfortable. Late in August we found the female incubating eggs. She wouldn't be flushed and, since the tower's height was then only eighteen metres, we decided to postpone work at this nest until the following year.

The female Square-tailed Kite perches beside her nest.

1975 saw us exceptionally busy along the Birds- ville Track and it was late in the season before we left to travel north. We drove through Birdsville Betoota, Windorah, Jundah and Longreach, and from there on the Capricorn Highway to Rockham- pton. The track from Birdsville to Windorah was the roughest we'd experienced until we hit the stretch on the Capricorn Highway between Jericho and Emerald. However, we still managed to cross the state from west to east in less than two days. From Rockhampton we turned south along the Bruce Highway until we reached the nest.

The Square-tails weren't using it. As we approached a large bird left from on or near the nest and flew away through the treetops, giving us only fleeting glimpses and no chance to positively identify it. The tower was not raised, although the nest was kept under observation for the better part of a day. The mystery bird did not return and the Square-tails were not sighted.

In October 1975 Cec Cameron telephoned say- ing the Square-tails were nesting again along the Wieambillia Creek. We left on the morning of 30 October and arrived at Cec's property, fifteen hundred kilometres away, that evening.

The following morning Cec showed us the nest. It was at a height of twenty-one metres in an Apple Box, close to the bank of the water-filled creek.

A female Square-tailed Kite with her well fledged chick.

It was on the fork of a thick horizontal bough, just like the nests we saw at Bororen and Port Lincoln. The loose sand at the foot of the tree frustrated our efforts with the tower, but Cec came to our rescue with his four-wheel-drive and effortlessly raised it. Our activity at the foot of the tree, the proximity of the hide at six metres from the nest, and Cec standing on the ladder just beneath it in full view of the brooding female, taking notes of her plumage, failed to flush her. In fact, after an initial threat-display - raising the nape and head feathers - she completely ignored all of us.

When the hide was fitted out with the cameras the female, as if on cue, delivered an incredible performance. She spread her wings to reveal her beautiful under-wing plumage, then appeared to solemnly bow to the photographer as he peered through the viewfinder filming her. She could be likened to a beautiful mannequin modelling a stunning new creation. We were lucky to capture this scene on film. As a curtain-raiser it was never repeated.

We worked this nest for a number of days. Throughout the night the female perched beside the nest in a vigil over her two almost fully-fledged chicks. Around 06:30 she flew off and returned within a few moments with a sprig of eucalyptus leaves which she spread on the nest; sometimes the chicks tried to help. About 08:00 the male arrived with the day's first meal. This was usually two fledgling birds, either Noisy Miners, or Little Friarbirds, plentiful in the area. He released them from his talons to the waiting chicks, stayed barely one or two seconds, then flew off to his usual perch in a nearby dead tree where he preened until ready to return hunting. The chicks swallowed the fledglings whole. About noon the female left the nest for an hour, probably to feed as she was never seen to do so while at the nest.

We first thought the male's short stay at the nest was caused by the nearness of the hide, but later realised that the male of several raptor species is not welcome near the nestlings.

In 1976 we worked at another nest on 'Rockwood' about ten kilometres north-west of the previous one. It was at the same height and was also sited on a large horizontal bough, although this time in a eucalypt beside a watercourse. There was one chick and the male brought one fledgling at a time. The adults behaviour was the same as at the previous nest, though, despite the relative proximity of the two nests, they were not the same pair.

When the male flew off to hunt, a commotion would go up throughout the area as the other birds heralded his approach. They knew him for what he was - a nest robber - and always attacked whenever he flew into their territory. On his route back, with a fledgling locked in his talons, the attacks ceased when he was about two hundred metres from the nest, and from there on in his flight went unmolested.

The tower at the 'Rockwood' nest, 1976.

Cec Cameron explained this as a case of mutual tolerance. Within a radius of about two hundred metres around the Kite's nest lay a non-combat zone. Within it, belligerancy between the male and the other birds ceased - they never attacked him and he never hunted them. Many species which the Square-tail regards as prey, nest within the zone, sometimes in the same tree. While the tolerance was mutual in this as well as at other species' nests, we have observed exceptions. The Black-breasted Buzzard tolerates other species nesting nearby with admirable forbearance, since they attack it at every opportunity - especially when on the nest.

The male Kite pauses briefly at the nest.

The male Kite arrives with prey for the brooding female.

While searching for the Red Goshawk in September 1977 we recalled the incident near Bororen when the unidentified bird flew from or near the Square-tails' nest and wondered if it could be the elusive Red Goshawk. We returned to investigate and found the Square-tails nesting at a height of seventeen metres nearby. The tower was erected with the hide four metres from the nest. One chick was being brooded by the female and she characteristically ignored us. She even obliged by standing on the edge of the nest as leaves were tied back beside her. The adults' behaviour was similar to those worked earlier, except the female fed on the nest from prey brought in by the male. This was probably normal because the chick was younger.

On one occasion an almost fully-fledged Friar Bird was brought in still alive. The female fed small pieces from it as it stuggled to escape. Some of these fledgling prey were so small, barely a mouthful for the chick, it hardly seemed worth the effort. The sight of this large raptor in its full predatory role, preying on a freshly-hatched chick

is most inglorious. Since during the breeding season their prey was entirely fledglings, we estimated the pair accounted for several hundred during that period alone. Nature has maintained a balance, however, the Square-tails are relatively rare and there is no shortage of the species on which they prey.

Although we were unable to work the Wieambillia pair that year, we did learn that they reared a chick despite the extremely dry conditions prevailing in the area.

In 1978, while travelling east along the Capricorn Highway near Anakie, we spotted a Square-tailed Kite carrying a stick and flying across the road ahead. After a short search we found a pair building a nest at a height of twenty-six metres in a tree close to the Highway. Typically they sited the nest on a horizontal bough and ignored us as they went about their work.

A few days later we found another nest at the Bororen site. It was the lowest we had seen, at a height of a mere sixteen metres. Some shots of the unconcerned female, standing with the eggs at

37

The female Kite remains on her nest as preparations are made for photography.

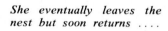
She eventually leaves the nest but soon returns ....

.... and settles down to brood once more.

Eggs of the Square-tailed Kite.

As it banks over its nest, this Kite could easily be confused with a Buzzard.

her feet, were taken from less than a metre away. The male ignored us as well, bringing prey for his mate. She always left the nest to feed, leaving him to take up his share of the incubating. This provided some evidence that his short stay at the nest in previous seasons, when the chicks were fledging, was not the result of the presence of the hide. The behaviour is better explained by his mate's distrust when chicks are in the nest.

There were now two major locations where we could study Square-tails: along the Wieambillia Creek near Cec's property 'Rockwood', and the site near Bororen. In 1979 and 1980 we made a number of journeys to both sites.

A portrait of the female Square-tailed Kite.

39

In November 1979 David Hollands accompanied us when we found another nest on the Wieambillia Creek. It was occupied by the female and a fully fledged juvenile. In July 1980, while searching for the Red Goshawk on Charlie's Creek, near Chinchilla, we came upon a pair of Square-tails building a nest at a height of twenty-two metres in an Apple Box. The nest was placed on a horizontal bough. Returning from a trip to the north in October 1980, we stopped to inspect the Bororen pair. They were using their 1977 nest and were rearing an almost fully fledged chick. This nest was approximately two hundred metres from the first found in 1974, and all of the nests for the intervening years were between these two. This nesting pattern was almost identical with the pair recorded by Cec Cameron on Wieambillia Creek from 1969.

The following day, while driving south, we stopped off to check our Wieambillia pair and found they were using their 1979 nest. They were feeding two chicks only a few days old. Therefore egglaying at this nest must have started about six weeks later than at the Bororen nest.

Beneath the tree, comfortably seated at a table, was that renowned avian enthusiast Roy Wheeler. He was keeping both birds under close observation while writing up his journal. We joined him to yarn for a few hours before retiring to the hospitality of the Cameron homestead. There we did our best to contribute toward the survival of the Queensland sugar industry by having a rum or two.

Cec phoned later to say the chicks left the nest on the Wieambillia Creek on 18 December, thus fledging was fifty nine days.

The female Kite feeds her almost fledged chick on a nestling bird.

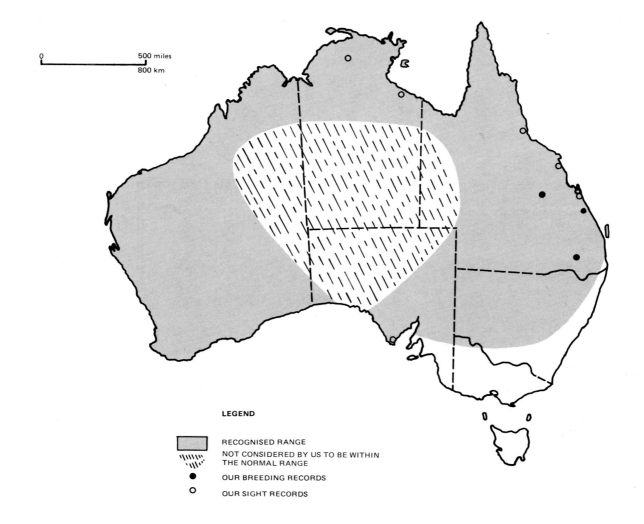

LEGEND

�earthgray	RECOGNISED RANGE
//////	NOT CONSIDERED BY US TO BE WITHIN THE NORMAL RANGE
●	OUR BREEDING RECORDS
○	OUR SIGHT RECORDS

0   500 miles
          800 km

SQUARE-TAILED KITE   *Lophoictinia isura*

lophos- crest (Gk); *ictinos-* kite (Gk); *isos-* equal or square (Gk); *oura-* tail (Gk)

LENGTH: 500 - 560mm. Female slightly larger than male.

WINGSPAN: Approximately 1300mm.

DISTRIBUTION: Rather rare, thinly scattered throughout Australia, in open forests and woodlands, absent from both heavily forested and treeless areas. Though generally considered nomadic and occuring in the inland, we have found it only in wooded areas near the coast or sub-interior where it is sedentary. Not found outside Australia.

VOICE: Usually silent; female's food call at the nest is a weak, high-pitched twitter.

PREY: Young birds, insects and reptiles. This kite does not appear to eat carrion. At nests we worked, the prey was entirely nestling birds.

NEST: A substantial structure of sticks and twigs, usually placed in a tall tree from 12 to 25 metres above ground. Though nests are reputed to be up to one metre in diameter, the nests we have seen were considerably smaller, being no more than 600mm in diameter; generally rather flat, with the centre cup well-lined with green leaves. With only one exception they were placed on a horizontal branch at heights of 16 to 22 metres above ground. Nests may be re-used or a new one built nearby.

EGGS: Usually two to three, 53 x 38mm. They are rounded ovals, buffy white without gloss, boldly spotted with red-brown and lavender. We have seen only one clutch in the field consisting of two eggs. Other nests contained two well-fledged chicks, one a half-fledged chick, one a fully-fledged chick and one two freshly-hatched chicks.

Eggs are laid from August to November, our records indicating egg-laying from late July to late September. At one nest the chick took its first flight at 59 days after hatching.

41

BLACK KITE
Milvus migrans

This species is one of the world's numerous birds of prey. It is widely distributed from Europe and North Africa, through southern Asia, to New Guinea and Australia. Throughout this range the various races are mostly migratory. The reasons for their movements are often obscure with some populations moving in opposite directions to others. In Australia, however, movement is related to the availability of food.

This propensity for a migratory or nomadic existence is reflected in the kites' scientific name *Milvus migrans* - *milvus* meaning 'a kite' and *migrans* 'wandering'. Throughout Australia the Black Kite is known as an Allied Kite, Kimberley Kite or a Kite-hawk, but is probably best-known as the Fork-tailed Kite.

Being extremely gregarious birds, Black Kites are seldom seen as individuals. Flocks of hundreds and, occasionally, thousands mill around killing centres, cattle yards and road camps throughout northern and inland Australia.

They are mostly scavengers feeding on carrion, refuse, injured birds and animals, or pirating prey from other birds. They also hunt small mammals, insects and reptiles, especially around grass fires or when plagues of such creatures occur. On one occasion we saw a Kite take a small live fish about 200mm in length from the flooded Cooper's Creek, but it wriggled free. In a Kite's nest along the Birdsville Track we found a house mouse and a 400mm long snake. Though bold when seeking food we found them quite wary at the nest where, with two notable exceptions, they showed no sign of aggression towards us. Avian intruders, however were dealt with unceremoniously.

The nest is a rough stick structure usually measuring 500 - 600mm in diameter - although some are reportedly larger. It is either unlined, or lined with practically any available material: rags, paper, fur, wool, dried skin, bark, leaves and, in cattle country, dung. Along the Strzelecki Creek one of the most commonly used lining materials were the dried sacks of caterpillars' nests. The height of the nest is dependent on the trees in the area. Along the Strzelecki Creek and the Birdsville Track, as well as other semi-desert areas of the Inland, nests were usually from a mere two to ten metres above ground. Along the Murray and Darling Rivers they were as high as thirty metres.

The Black Kite is generally an opportunity breeder, nesting whenever food supply is adequate, although in the southern limits of its range breeding is often in the spring. The size of the clutch is also affected by prevailing conditions. During the 1974-76 rat irruption in the Inland most nests we checked contained three or four eggs. In

A small clutch of well marked Black Kite eggs.

subsequent years in the Inland and in northern Victoria, when conditions were less favourable, clutches more often consisted of just two eggs.

During the rat irruption the Kites built up to tremendous numbers. By August 1976 the rat hordes had been decimated by their predators, and the Kites dispersed towards the coasts to the north, east and south. Some, especially young birds, left it too late to get out and they died of starvation. We picked up numerous birds at the point of death.

The Black Kite is generally uncommon in all of Australia's southern regions. Until late 1976 we found only a few nesting in tall timber along the Murray River, far beyond the scope of our towers. However, they appear periodically after Inland irruptions - the previous such influx being in 1952. In late 1976 thousands appeared in the Victorian Mallee, and in the spring of 1977 we saw them nesting in greater numbers than ever before. They bred there for several years.

Because we found the Kite so common in the Inland we tended to ignore it in favour of rarer species. Our first photographs of it were obtained quite by accident. In July 1975, together with David Hollands, we camped near Damperanie Well on the Birdsville Track, having seen several pairs of Black Falcons in the surrounding area. We erected a tower at a Falcon's nest twenty kilometres to the north, and were searching for a second nest in a group of trees not far from our camp. David climbed to a likely-looking nest to find three small chicks which, he declared, were 'Black Falcons, or I'm a Chinaman'. We set up a tower and retired a short distance to await the falcon's return.

Shortly afterwards a Black Kite twice passed low over the nest, almost landing on the second pass. Puzzled over the Falcon's failure to drive off the intruding Kite we ran towards the nest-tree, shouting and waving. The Kite flew away to a nearby tree and we returned to our vantage point. Almost immediately it made another low pass over the nest, banked and alighted on it. Again we drove it off, then checked the nest. We were greatly relieved to find the chicks unharmed.

Believing the proximity of the tower was deterring the Falcons from protecting their brood, we hastily lowered it and resolved to work the nest to the north, where the Falcons had readily accepted the tower. David was installed in the hide and we returned to the camp. We decided to raise the tower at the second nest and immediately occupy the hide to protect the chicks.

Within five minutes a Black Kite landed with a large native rat which it proceeded to feed to the chicks. Suddenly everything was clear. We were at the wrong nest! Later that day we found the Black Falcon's nest close by. When David was relieved from his hide that evening, he received much good-natured bantering about his knowledge of birds - and his pedigree.

A female Black Kite on her nest in a coolabah along the Birdsville Track.

A black Kite in flight.

Although the Kites dispersed from the Birdsville Track area late in 1976, they were still numerous along the Strzelecki Creek. They were not breeding *en masse*, but throughout the year, with periodic upsurges in breeding activity when local conditions were favourable. During an extremely hot period in October 1977, when we were near one of the very few waterholes still containing water, hundreds of Kites would arrive in mid-afternoon and perch in the Coolabahs near the water's edge. They would remain there until evening with their wings spread and beaks agape, blackening the tree-tops like a huge colony of flying foxes. On days when the heat was less extreme they hunted in the surrounding treeless sandhills. At dusk they arrived in thousands to roost in trees along the creek.

In November 1977 we found a pair nesting six metres above ground in a belar tree on the edge of a wheat farm, only twenty kilometres from home. The nest contained three eggs. On 28 November they hatched and we began our photography nine days later.

The chicks were brooded constantly by the female which only left the nest to remove bones or unsuitable fragments of carcass. The male brought all prey to the nest. Apart from a few grasshoppers, it was predominantly rabbits, probably picked up from roadsides where they had been killed by motor vehicles. They were either kittens and small rabbits, or portions of fully-grown ones, since this kite is incapable of flying with a large rabbit. The female always uttered a few loud quavering whistles when she saw the male approaching with prey. He carried all prey in his talons, although after alighting he sometimes took it in his bill and passed it to his mate.

He seldom stayed long at the nest. We were unsure if that was normal behaviour or the result of the close proximity of the hide. Apparently it is unusual for the male to feed the chicks, but on one occasion he fed a large grasshopper to a chick before the female could intervene. On several other occasions he seemed about to feed the chicks when she darted across the nest, took the prey from him and fed them herself.

As the chicks developed she stopped day-brooding, unless it was hot, when she shaded them with her body and outstretched wings. On

The male Kite helps feed the three small chicks.

44

Male Black Kite.

Female Black Kite.

The female Kite feeds one of her chicks.

When the chicks were five weeks old and fully-fledged, the tree was climbed to photograph them in the nest at close range. Surprisingly the female kite became extremely agressive. Emitting a shrill staccato **'kee-ki-ki-ki'** she swooped within a metre of the intruder. This was our first experience of agressive behaviour from a Black Kite, despite having climbed scores of trees inspecting nests at all stages of the breeding cycle. It was not the last. Two years later we were again attacked by the species. A Kite was defending an empty nest it had just acquired. Some weeks later, when there were eggs in the nest, our intrusion resulted in the unwary photographer being struck on the head, drawing blood.

one extremely hot day she shaded them for hours and ignored the half-rabbit the male had brought in. Even in the shadow of her body the chicks were very restless and her concern for them was obvious. Her stoicism was so profound that, despite the sauna-like conditions inside the hide, the photographer was, momentarily, emotionally overwhelmed by wonder and respect.

The chicks showed no aggression toward each other, each getting its fair share of food without fuss. Between feeds they showed little animation, even when three weeks old. They quietly passed the time in the bottom of the nest, or gently nibbled at their downy breasts. Unlike most other young birds they did not back to the edge of the nest to defecate. Instead they lowered their heads, raised their rear-ends and ejected howitzer-like from the cup of the nest. So well did they perform, the nest remained clean, inside and out.

The first chick left the nest at thirty-eight days and the others at forty-one. They were still being fed on the nest on the forty-second day when our observations ceased.

Our study of this pair revealed an interesting pattern of behaviour. We were using a red Ford Falcon utility which the birds grew to recognize. Whenever it made a right turn off the main road, a kilometre from the nest, the adults swiftly flew away. Any other vehicle could drive past the nest-tree without disturbing them. Their vigilance and quick response always alerted the occupant of the hide that his seeing-out party was within two or three kilometres. If both birds were on the nest the male left first, while the female waited until the utility made that right turn a kilometre away. Our conviction that the birds associated the red utility with the photographers who invaded their privacy, was strengthened when the seeing-out party used a different vehicle. The female did not leave the nest until the vehicle stopped at the nest site.

The five week old Kite chicks.

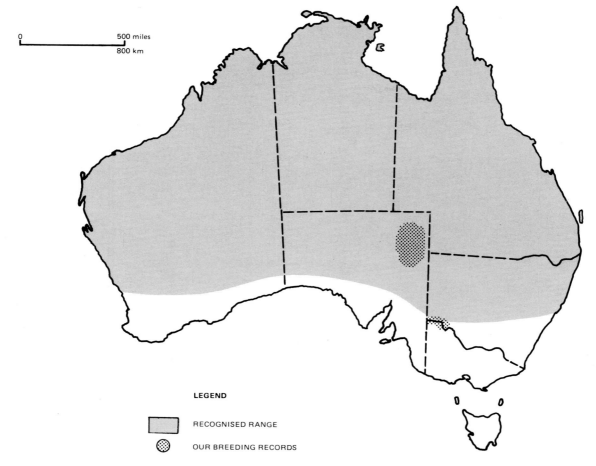

LEGEND

RECOGNISED RANGE

OUR BREEDING RECORDS

BLACK KITE *Milvus migrans*

milvus- kite (L); *migrans*- wandering (L)

OTHER NAMES: Fork-tailed Kite; Allied Kite; Kimberley Kite; Kite-hawk.

LENGTH: 480 - 550mm. Female slightly the larger.

WINGSPAN: Approximately 1200mm.

DISTRIBUTION: Very common in northern and inland Australia; nomadic. Huge flocks congregate around inland killing centres, stockyards and refuse dumps. Usually uncommon in the south but becomes numerous after big build-up of numbers in the interior during rabbit and rat irruptions. The Australian subspecies *M.m.affinis* extends through New Guinea, Timor and Sulawesi. Other subspecies range through Asia, Europe and Africa.

VOICE: A high-pitched quavering **'kwee-errr'**; a staccato **'keee-ki-ki-ki'**.

PREY: Small mammals, reptiles, grasshoppers, frogs, disabled birds, stranded fish. Garbage and carrion is also a major food source. Rabbit formed the bulk of the diet at nests we worked, the only other items recorded being rats and a large insect.

NEST: A rough structure of sticks, usually 500 - 600mm in diameter, occasionally larger. The nest may be lined with any of the vast range of material available to them. We have noted rags, paper, fur, wool, dung (in cattle country), dried skin, bark and occasionally leaves. Some remain unlined. In several nests the dried sacks of caterpillars nests had been used for nest lining. In the interior we found nests as low as two metres above ground, while along the Murray River they nest at heights up to thirty metres.

EGGS: Usually two or three, 51 x 42mm, dull white, sometimes almost unmarked, usually sparingly spotted and streaked with red-brown and underlying purplish-grey. During 1974 to 1976 most clutches were of three eggs. Our records since then show that, of 25 clutches - 5 contained three eggs, 18 had two eggs, and 2 were single-egg clutches.

Breeding will take place at any time of the year, according to food supply, although in southern Australia breeding is more often in the Spring.

The incubation period appears to be about 35 days and chicks fledge in 38 - 42 days.

WHISTLING KITE
Haliastur sphenurus

This species, sometimes known as the Whistling Eagle or Carrion Hawk - but more correctly the Whistling Kite, is found throughout Australia. It prefers open forests and woodlands near lakes, swamps, timbered watercourses and coasts. Less common in south-west Australia, and a rare visitor to Tasmania, it is found in New Guinea, the Solomon Islands and New Caledonia. We found it in most areas we visited. Along the Murray River we often found them nesting above thirty metres.

The nest is a bulky structure of sticks, lined with green leaves. It is usually in a dominating tree, often over or near water. The same nest may be used year after year and becomes progressively larger - we have seen nests over one metre deep. The breeding season varies according to the area and food supplies. In the inland breeding may occur at any time of the year, while in south-eastern Australia, August to October is more usual. We found a pair brooding eggs in February on Lake Victoria, in south-west New South Wales, and another with almost fully fledged chicks along

48

the Strzelecki Creek in April.

The distinctive whistle, a long shrill note followed by four to six shorter ones with a rising inflection, leaves no doubt as to the bird's identity. In flight the long rounded tail differentiates the Whistling Kite from the Little Eagle or the Black Kite. We noted a large variation in plumage, partly due to maturity - the young generally being darker.

The Whistling Kite in flight.

In September 1975 we set up at a nest in a Coolabah in a dry creek bed along the Birdsville Track. The loose sand was tricky and we were almost bogged a number of times. The nest was about ten metres above the ground and it contained one chick about three weeks old. The parent birds remained wary of us. A gusty northerly hadn't helped them settle down since it caused the tower and hide to move. Worse was to come. Around 17:00 the wind dropped completely and an eery, ominous silence settled over the creek bed.

Far away to the south across the gibber someone appeared to be approaching very fast, judging by the huge cloud of dust being thrown up. Then the realization grew it wasn't a vehicle at all, but a change in wind direction and velocity. There was barely enough time to grab the camera gear and vacate the hide. The storm struck as the ground was reached. One guy rope slipped easily from the sandy creek bed and the tower swung wildly, straining against the other guy tied to the nest-tree. Together we tried to pull the tower back to vertical against the wind. When that failed we tied the loose guy to the nest tree, found what cover we could and waited it out. Soon the worst was over and we managed to get everything ship-shape before dark.

The next morning the hide was entered before daylight. The parent birds were about early calling to each other. They remained wary of the nest until 09:45 when the female alighted. In a bid to allay her suspicion of the hide no shots were taken. The strategy worked and in a short while the male alighted with prey, probably a pipit and the female fed the chick. We would have stayed a day or two longer at the nest, but unfortunately time did not permit.

A pair of Whistling Kites with their half fledged chick.

A typical clutch of Whistling Kite eggs.

Kite chicks soon after beginning to fledge.

Fully fledged Kite chicks.

The adults feed on anything from insects to carrion. With the chicks the diet is more particular. We never saw carrion fed to them, the closest being the fresh placenta from a cow - perhaps a greater delicacy. Generally they seemed to feed on other small birds, rats and rabbits. Small carp that sometimes mass in large numbers around water regulators within our irrigation district will attract the kites in flocks.

We worked them again in the Cobblers near Montecollina Bore in June 1979, where a pair nested amid the Letter-wing colony. The adults were attentive when the chick was small and we obtained some good feeding sequences. There was also a humourous encounter between a scavenging Raven and the lone chick. The Raven consumed odd scraps in the cup of the nest while the chick cowered on the outer edge. After a short time the chick appeared to lose some of its initial fear of the Raven and, like David to Goliath, turned to face it. The chick made a lunge and, although it was only a feint, it was enough for the startled Raven. It took off hurriedly and did not return.

Soon afterwards the sound of bones being crunched came from below. A peep revealed a dingo feeding on the remains of rabbit carcasses dropped from the nest. The adult birds stooped at the feeding dingo, with little result. The dingo ignored them and, after consuming all the rabbit

The female Kite comes to her nest with prey.

She feeds her chick from a Bearded Dragon.

The juvelile Kite alights after its first flight.

remains, it trotted off unperturbed by the two kites stooping to within a few centimetres of its back.

We left the nest for a fortnight with the tower in place. When we returned and occupied the hide we were unable to get the adults to the nest. We tried every stratagem we knew, but to no avail. We removed the tower and left it at 'Merty Merty' Station, one hundred kilometres to the north, while we searched for other species to film. We felt sure that, had we not been absent for a time, we would have had no problems. We found this was a common reaction among some species:

they would come to accept regular comings and goings, but would become wary after being undisturbed for a time, despite the continued presence of the tower.

On returning to the area in October we found the kites had bred again in the same nest and were feeding two small chicks. Although this was the first time we recorded this species breeding twice in the one year we were not unduly surprised. They, like many inland birds, are opportune breeders and, with rabbits and other prey in plague proportions, conditions were ideal.

LEGEND

◻ RECOGNISED RANGE

⊛ OUR BREEDING RECORDS

WHISTLING KITE   *Haliastur sphenurus*

hals- sea (Gk); *astur-* goshawk (L); *sphen-* wedge (Gk); *oura-* tail (Gk).

OTHER NAMES: Whistling Eagle; Carrion Hawk.

LENGTH: 510 - 580mm. Female slightly larger than the male.

WINGSPAN: Approximately 1200mm.

DISTRIBUTION: Common throughout most of Australia, particularly in open forests and woodlands near lakes, swamps, timbered watercourses and coasts. Some birds are sedentary while others - particularly in the interior - are nomadic. Less common in the south-west and a rare visitor to Tasmania. Also found in New Guinea, the Solomon Islands and New Caledonia.

VOICE: A loud, drawn-out descending whistle, followed by a series of short whistles of rising inflection. The alarm call, when disturbed at the nest, is a mournful **'kairr'**.

NEST: A bulky structure of sticks lined with green leaves, usually in a dominating tree and often over or near water. Nests are used year after year and become progressively larger. We saw nests over one metre deep, although the nest diameter seldom exceeds 600 - 700mm.

EGGS: Two eggs usually form the clutch, although three-egg clutches are fairly common. Measuring 57 x 44mm they are coarse, pale bluish-white ovals, usually sparsely marked with red-brown and purple; occasionally they are well-marked. Of 22 clutches recorded by us, 18 contained two eggs and the balance three eggs. Egg-laying may occur at any time of the year according to food supplies - this is particularly so in inland areas. In most of south-eastern Australia August to October is the normal period. Our records show egg-laying in every month.

PREY: This kite feeds largely on carrion of all types, but will also take live prey such as birds, small mammals, lizards, fish and insects and their larvae. We have seen birds, rabbits, rats and lizards fed to the chicks.

BLACK-BREASTED BUZZARD
Hamirostra melanosternon

Although it is not uncommon for birds to be renamed as more becomes known about them, few have proved more enigmatic than the Black-breasted Buzzard. Though referred to as a Buzzard from the first, it was soon realised it was not a true Buzzard of the Buteo genus and was placed in a genus of its own - *Hamirostra*, from the Latin *hamus*, meaning 'hook' and *rostrum*, 'bill'. *Melanosternon* comes from the Greek *melas* - black and *sternon* - breast.

This arrangement proved satisfactory until recent years, when, because of its apparent closer relationship to the kites, it was renamed Black-breasted Buzzard Kite and later Black-breasted Kite. However the widespread acceptance of the original name prevailed and today the recommended name is once again Black-breasted Buzzard. To us it will always be the ''Buzzard''.

It is a rather rare species, inhabiting the open woodlands and savannas of northern and inland Australia. In arid areas it is found along timbered watercourses. Though usually sedentary, harsh conditions can force some movement of the species.

The Black-breasted Buzzard in flight.

Gould referred to this species using a stone to break open Emu eggs to get at the contents. The story is said to have originated from the Aboriginals and has never been recorded as having been witnessed by white settlers, although it has been given some credence by reports of stones being found in Emu nests along with broken eggs robbed of their contents. There have also been reports of emu and bustard egg-shell being found in the Buzzards' nests. In more recent times photographs showing vultures in Africa using a stone to break ostrich eggs have also tended to add credence to the story.

Mr. R. H. Bennett, while at Yandembah Station in the Lachlan District of New South Wales, describes in 1881 how rare this bird had become, whereas it was quite abundant in earlier years.[8] He blamed the shyness of the bird and perhaps the removal of the larger trees. He, like many of his contemporaries, never mentioned the possibility that the excessive taking of eggs and skins could have been a reason for their decline in numbers. The species is not a shy bird as we know it in the inland areas, and it is not always a high nesting one. We've worked them on nests below seven metres where much higher trees abound. Anyone who has read A. J. North's **Catalogue of Birds of the Australian Museum** must, like us, be appalled at the recorded excesses of egg and skin collectors in those days.

In flight this raptor is generally quite easy to distinguish from any other because of the white 'windows' in the wings near the base of the primaries. The white patch is also on the upper surface. The only time there could be any confusion in identity would be with a light phase of the species, or with an immature bird, when it might be mistaken for the Square-tailed Kite or the Little Eagle, as the 'windows' don't contrast to the same extent. When perched it could be mistaken for a Wedge-tailed Eagle as the white windows do not show unless the wings are extended. If the legs are visible there shouldn't be any confusion as they are clean, not trousered as with the eagle. When soaring the wings are carried in a higher 'V' relative to that of the eagle. The very short tail of the kite gives it a huge bat-like appearance when in flight.

Our first sighting of this raptor was along the Birdsville Track in August 1974. A pair were flying westward at a great height at the time but there was no way they could be mistakenly identified. We later saw one perched beside a nest on Marrapinna Station one hundred and sixty kilometres north of Broken Hill, New South Wales. The nest was a large, rather flat platform of sticks, at a height of sixteen metres in a large river gum on the edge of a waterhole on Noontherrungie Creek. We inspected the nest and found it well-lined with fresh eucalypt leaves and apparently ready for egg laying. A further inspection three weeks later found the nest the same, freshly lined with leaves but still without eggs. It was not used

that season nor the next, although the one Buzzard was always nearby. As we at no time saw two Buzzards we were forced to the conclusion that it had lost its mate, and, being near the southern limits of its normal range, had been unable to find another. Egg collectors had been known to operate in the area for some years and they in no way help the survival of our rarer species.

After extensive searching, a pair were found nest-building along the Strzelecki Creek south of Innamincka in the north-east of South Australia, on 16 August 1976. We kept them under observation from a discreet distance and they were seen to mate in a tree next to the nest-tree. The latter was a Coolabah, and the nest was at a height of twelve metres.

Almost a month later we again visited the nest and found them incubating two eggs. Both birds were on the nest but we can't be certain that incubation is shared, but suspect that it is. They kept a very good lookout as one bird was seen to leave the nest when we were still a long way off. The other crouched low in the nest till we were directly below it before it also flew. This we found to be the usual pattern of behaviour except that the remaining bird was sometimes hard to flush. While we were inspecting the nest both birds circled the nest-tree voicing alarm calls. The call was a short sharp yelp repeated at intervals of one or two seconds. It was reminiscent of the bark of some small lap-dogs. Both birds returned to the nest as soon as we were a few hundred metres away.

On approaching the nest on our next visit we found the Buzzards more vociferous than usual and we were not surprised to find a freshly-hatched chick. The second egg did not hatch and it was found later to be addled. The tower was erected well back from the nest on 14 October and left to allow the birds to get used to it. It was moved closer two days later and the hide entered.

The brooding and hunting was fairly evenly divided between the adults for the first few days. This pair were similar in size and we were unable to decide which was male or female. The hunting, feeding and brooding ritual was intriguing, and as far as we can ascertain unknown among other raptors anywhere in the world. The non-brooding bird left the nest and soon returned with a rabbit which it quickly plucked and/or skinned on the nest. It then attempted to feed the brooding bird with small pieces of flesh in much the same manner that a chick is fed, but it refused to take the proffered pieces. The refusal seemed to worry the host bird and it would eat the piece after several profferings had been refused. Increasingly smaller pieces were proffered but to no avail. The cine camera noise was obviously upsetting the brooding bird, so it was stopped long enough for it to accept a tiny morsel, but dropped it in the nest as the camera was restarted. The morsel was retrieved by the other and proffered again with the same refusal. The concern expressed on the one

The brooding Buzzard is fed on the nest by its mate.

After the brooding bird is fed it moves to the edge of the nest before accepting more prey . . . .

. . . . which is then relayed to the chick.

A Buzzard alights above its nest.

hand and the sullen refusal to eat by the other was very real and no figment of our imagination. Many people who have viewed the film have commented on it and one viewer said twelve months later that the concern so evident in the host bird was one of the most beautiful things she had ever witnessed. Filming was halted long enough to allow the brooding bird to settle down and eat the proffered food. After its appetite was sated it rose slowly and carefully from the chick and moved a little toward the nest edge from where the tiny morsels were accepted and relayed to the chick. After it had been fed the adults changed roles. The bird that had been brooding went hunting while the other brooded the chick. This was the general behaviour pattern for the first few days with only slight variations. Some times both attempted to feed the chick at the same time but the brooding bird was always fed before the chick. Many male birds among other species bring prey to or near the nest but the female always feeds herself on or near the nest.

Rabbits were the predominant prey brought to this nest but occasionally Little Crows, Galahs and Kestrels, both adult and fledgling figured in their diet.

All manner of birds attacked the Buzzards in flight as well as on the nest. Small birds such as Wood Swallows, Magpie-larks and Wagtails attacked them on the nest, while Corvids, Magpies

From its vantage point the Buzzard scans the surrounding country.

56

and Black Kites were their main antagonists in the air. A Black Kite was seen to lock talons with one of this pair and they tumbled earthward to what appeared must be certain death, but they parted to cheat it by a few metres. A Black Falcon nesting about half a kilometre away joined in the harassment of them occasionally.

This raptor, like many of our larger ones, lives in harmony with Zebra Finches which quite often nest in the substructure of raptor nests. We have been unable to discover any benefits arising from this common practice for either species, but there no doubt is a reason for it. Perhaps the finches find the raptors' nest as protective as the thorn bushes they often nest in. On one occasion a finch hopping in and out of the entrance of its nest well back in the substructure of the buzzards', was attacked by another bird, possibly a Pied Butcherbird. Positive identification was not possible, so fast was the action and viewed as it was, through the confines of the camera's telephoto lens. The attacker missed, due no doubt to quick evasive action of the finch which shot into the entrance to its nest and did not emerge for quite a long time.

The behaviour of the Buzzard is much akin to that of the Square-tailed Kite which it also resembles in size and plumage to some degree. Both species are nest robbers, although the Buzzard's range of prey is far greater than the other. Both sit tight during brooding, but the Square-tail more consistently refuses to be flushed by humans. Each generally lay two eggs and occasionally rear two chicks. The plumage of the light phase Buzzard and that of the Square-tail is very similar. The chicks of each would be hard to differentiate during the fledging period.

The Buzzard chick was downy white for the first two weeks and gradually thereafter took on a reddish-brown plumage. At one month the head, nape, wings and tail were feathered, but the breast and back were still downy. At forty-five days we were surprised to find the chick missing. We learned later it had been taken by humans.

We found four other nests in 1976 though, of these, only at one was breeding success confirmed, two chicks being reared.

Returning to the Strzelecki in April 1977 we found a group of four birds close to the latter nest.

From that it would appear that the species is, to some extent at least, sedentary, and the juveniles are dependent on their parents for some months after fledging. These two would have left their nest about the end of November.

In June we found the nest of that group completely demolished. Apparently it had blown down, and judging by the large accumulation of sticks around the base of the tree, this must have happened most years. Not a stick remained in the tree on this occasion, but by the end of August it had been rebuilt in exactly the same position, and the female was occupying it. It was easy to identify the sex of this pair as the female was noticeably larger and she was of the light phase.

A typical clutch of Buzzard eggs.

An interesting point about this nest was its immunity to egg collectors, being at the top of a dead Coolabah and on incredibly thin branches. It was the only one so placed of all we've seen, as they are usually built on more substantial branches and quite often only half way up the tree. We've since learned that an egg collector robbed all other nests in the area in a single season. Another interesting feature is the species' preference for nesting in dead trees. Almost all nests that we have found to date have been in dead trees or the dead part of a growing one.

The Buzzards were still on eggs when we erected the tower on this nest and the first chick hatched on 12 October, the same date that the chick had hatched in the nest we worked in 1976. The second did not hatch till five days later, and we didn't expect it to survive as it often missed out at feeding time. When small fledgling birds were brought in the larger chick ate them as quickly as

57   *The tower at the Mundibarcooloo Buzzards' nest.*

The Buzzard shades her chicks.

A large Goanna provides the chicks with several meals.

they could be fed to it. The only time the smaller got anything was when rabbit was on the menu, and the older chick's appetite would be sated before the rabbit was finished. The younger also came in for some severe pecking in the absence of the adults, but luckily that was not often, as the adults changed roles at the nest on most occasions.

Back at the nest a month later, we were pleasantly surprised to find both chicks doing well, although there was still a marked difference in size and maturity. Another surprise was the aggressiveness of the younger one. It was the first to be fed and the elder and larger waited patiently till the other's appetite was sated. It brought back memories of the school bully's demeanor after he'd been thrashed by a lad half his size. A large goanna provided them with ample food for a day.

The skin was very tough going by the exertion needed by the adult to tear it into manageable pieces.

The year 1977 had not been a good one for the species. They were still around in numbers comparable to the previous one but were not breeding. The one described was the only one found whereas in 1976 we monitored five nests. Incubation appeared to be about forty days.

A pair was sighted by us in Victoria not far from our home in 1977. It was the first time we'd seen them in this state and one of the few sightings recorded for the state. Like many other raptors, their numbers had built up during the high rainfall years in the centre, and the harsh conditions that followed forced them out to seek prey elsewhere.

The five days difference in the chicks ages is apparent.

Lizards are popular in the Buzzards' diet.

Mundibarcooloo Waterhole after heavy rains.

Our camp with the A. B. C. film crew, in the bed of Mundibarcooloo Waterhole, twelve months later.

A visit to the area in May 1978 found conditions extremely harsh as there had only been a few milimetres of rain over the preceding twelve months. All surface water had dried up except a little green "soup" in the Mundibarcooloo waterhole. The pair that had reared two chicks in each of the two preceding years were still around and would probably breed there again if conditions only improved a little. A flush down the creek would have done wonders for the area. The Cooper had substantial water in many places as a result of heavy rains in central Queensland during the preceding wet season. Even so, its waters hadn't risen high enough to flow into the Strzelecki at Innamincka.

The last time the Strzelecki had flowed was in 1974 when it was a raging torrent. Heavy local rains, combined with a flooded Cooper, caused a lot of damage to oil rigs in the area. We found drums of fuel still up trees two years after the flood. Ted Riek's family at 'Merty Merty' had been lifted out by helicopter and were unable to return for five months. It's a tough area from which to carve a living but its harshness didn't have any ill effects on the longevity of at least some of its inhabitants. We came across the grave of one A. Paterson aged ninety-four in 1917. Another pioneer, Mary Scobie, had died in her ninety-ninth year. Grandson Ross Scobie now worked with Ted and George Riek. This area had been settled and deserted before the latter had taken it up.

Merty Merty homestead surrounded by the 1974 floodwaters.

A further visit in August found a pair of Buzzards circling the nest near 'Merty Merty' homestead. We went on to Innamincka and crossed the causeway over the Cooper. There was a flow of fifteen centimetres which presented no problems, but there was a potential one if the creek should rise while we were on the northern side, as we would have to make a detour of several hundred kilometres to get home. As a consequence we limited our search to a few days. The Grey Falcon was our main objective as we'd seen a pair there in May. They weren't sighted, but a pair of Buzzards were observed occupying a nest overhanging water in the north-west branch of the Cooper.

The grave of an early settler, on the banks of the Strzelecki Creek.

The area had received some rain since our May visit and a carpet of predominantly yellow flowers stretched to the horizon in most places. Occasionally there was a mixture of blue and white amidst the yellow. We'd never seen such a colourful display before, even in the wet years. Ted explained that the timing of the rain was more important than the quantity as to which species of herbage germinated.

As we left 'Merty Merty' on our way home we noticed the Buzzards occupying the nest near the homestead. It was a mixed pair, a dark male and a light phase female. We wondered if it was the pair from the nest at the Mundibarcooloo waterhole. That nest had not been repaired and there was no sign of them in that area which was only a few kilometres downstream from the 'Merty Merty' nest.

We went back in October but found the nest deserted. Whether egg collectors had been around we couldn't be certain but when we found a similar pair back at the Mundibarcooloo nest we thought it highly possible that they had been robbed of their eggs and were trying for a second clutch. The female was occupying the nest but when we inspected it from the tower next day it was without eggs. Ted sent a telegram a few weeks later saying he'd inspected the nest site twice and found no activity.

Heavy rains in January 1979, with good follow-up rains, provided ideal conditions for many raptors in the inland, and we made many trips to the Strzelecki area.

On one such trip, early in August, we were fortunate to witness the courtship flight of a pair of Buzzards. We were travelling through Sturt National Park, in the extreme north-west of New South Wales, and had stopped to refuel our vehicle on top of one of the seemingly endless sandhills that stretch from Fort Grey through to the Strzelecki Creek. About one kilometre back, we had crossed a dry watercourse timbered with small coolabahs, but the area around us was completely treeless, so we were surprised to see a pair of Buzzards circling overhead. The birds were soaring slowly and silently in fairly parallel

A Buzzard chick begins to hatch.

flight. After about one minute, one bird drifted about 100 metres from its mate, slowly gaining height until it was perhaps 20 metres higher. From this position it swooped towards its mate at moderate speed. The second bird rolled onto its back and presented talons, though the two birds did not touch.

The display was repeated three times, the second and third encounters coming in rapid succession. This seemed to be the peak of activity,

with the birds appearing to touch talons on the third pass. They were drifting back towards the timbered area as they made their fourth pass - a rather half-hearted affair, with the lower bird only rolling a little as its mate passed several metres above it.

Although we were sure the Buzzards would have nested nearby, we didn't find the time to make a search. However, we found and monitored seven Buzzard nests along the Strzelecki, six of which had two eggs and the seventh one. Only the Mundibarcooloo pair reared two chicks, as they had in 1976 and 1977, and the others one each. When we left the area in November all were fledged or close to it, so we could confidently say eight chicks were reared from the seven nests. We did a little filming at three of them.

At one nest on the north side of 'Merty Merty' homestead the female appeared to do all the hunting. She would 'bark' orders at the male but the extent of his efforts was a few circuits of the nest-tree before alighting again. He was without doubt a loafer, and David, who was working with us, endorsed our observations.

Sixty days later, the chick prepares for its first flight.

61

A typical Buzzards' nest, the third we worked in 1979.

During our vigil the female brought in a partly fledged Black Kite chick. When released, it walked past the male and stood beside the Buzzard chick. All just watched the forlorn-looking chick for perhaps a minute before the female reached forward, picked it up in her bill, placed a talon over it and proceeded to tear off small pieces to feed the chick. There was no sign or sound of protest from the victim.

By contrast David described how a partly fledged Kestrel was brought in next day and released in the same manner. It was not prepared to be eaten without a fight. It lay on its back, screeching and striking at any head coming close to it. The Buzzard chick made a tentative peck at it and had its face raked by a talon. It backed off to watch from the sidelines. The female eventually grasped the still screeching and squirming Kestrel in a talon and calmly tore off morsels to feed her chick. It was nature in the raw.

At the Mundibarcooloo nest we recorded the two fully fledged young being fed a Bearded lizard. They were apparently popular in the Buzzard's diet as we found the remains, mostly the rough horny skin of them, hanging on the outer edges of most nests.

1979 was a good year for the cattlemen as well as for the birds. The heavy rains of January were followed by further good falls at the optimum times. The cattle in the semi-desert areas do far better in good seasons than those in the better rainfall areas of northern Australia. We saw cattle that were virtually just bags of bones in early 1978, but were being sold off in prime condition a few months after the drought broke. Even so, there are far more dry than wet years and it's in the dry ones that the delicately balanced ecology of the area can be irreparably damaged by over-grazing.

We would like to see at least the flood plain of the Strzelecki declared a National Park. It could be acquired relatively cheaply, and if properly administered, would make a wonderful refuge for some of our rarer avian species. At present there are two major threats to the effectiveness of this area as such a refuge. The first is its vulnerability to egg collectors - a very real threat to some species, and one that will only be overcome with constant and dilligent policing of the area.

The second, and wider reaching threat arises from the building of a levee which prevents normal flows of water down the Strzelecki. Unless floodwaters from the Cooper are allowed to flow naturally into the Strzelecki as they did in the past, the whole area could soon become a wasteland.

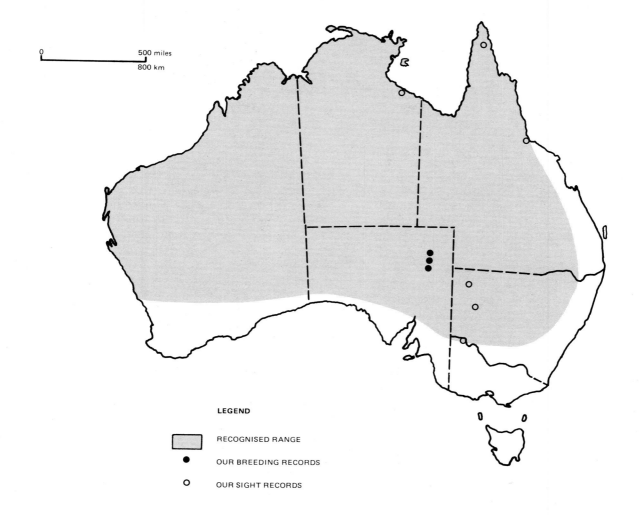

LEGEND

	RECOGNISED RANGE
●	OUR BREEDING RECORDS
○	OUR SIGHT RECORDS

BLACK-BREASTED BUZZARD
Hamirostra melanosternon

hamus- hook (L); *rostrum*- bill (L); *melas*- black (Gk); *sternon*- breast (Gk).

OTHER NAMES: Black-breasted Kite; Black-breasted Buzzard Kite.

LENGTH: 530 - 610mm. Female slightly larger than the male.

WINGSPAN: Approximately 1500mm.

DISTRIBUTION: Rather rare, thinly scattered through open woodlands and savannas of northern and inland Australia; absent from densely forested country. In more arid regions it is found along timbered watercourses where it may occasionally become locally numerous. Generally sedentary, but harsh conditions probably force some movement.

VOICE: Short, rather thin whistling scream. Also a sharp **'chip-chip'** and a short rasping yelp.

PREY: Rabbits (most common prey item), reptiles, adult and nestling birds, lizards (remains at most nests), a goanna, nestling Kestrels, Ravens, Black Kites, adult Galahs, Miners and several unidentified passerines.

NEST: A large, coarse, rather flat platform of sticks with a centre cup lined with green leaves, placed in the fork of a tree, 6 - 20 metres above ground. Nests we studied varied from 700 - 1200mm in diameter and were 380 - 530mm deep, with a centre cup, often quite deep, 350 - 450mm in diameter. Most were placed on a horizontal fork in a dead or thinly foliaged tree, 6 - 11 metres above ground.

EGGS: Usually two, 62 x 48mm. They are rounded ovals, coarse and without gloss, white or buffy-white, boldly blotched with red-brown, chocolate and light purple. We recorded eleven clutches: one single egg and the rest containing two eggs. Eggs are laid from July to November; our records show egg-laying from early August to late September. The incubation period is about forty days and fledging about sixty.

BLACK-BREASTED BUZZARDS
REARING AND PREYING ON KESTRELS

At 03:30 on 16 September 1976 we left home for the Strzelecki Creek. The weather had been dry for some time so we took the shortest route - travelling via Broken Hill, Tibooburra, Fort Grey and then along the New South Wales - Queensland border to Cameron's Corner where the Queensland, New South Wales and South Australian borders meet. The border is marked by a two and a half metre high netting fence known as the 'Dog Fence'. It was built in 1887 to protect sheep from the dingo, a native dog common north of the fence-line. The track along the fence was suitable for four-wheel-drive vehicles only because it crosses sand-dunes which run at right-angles to the fence. We managed to cross these by racing at them with the accelerator pressed to the floor, hoping there wasn't anybody else doing the same thing from the other side. On one occasion this happened. Fortunately the other driver had tied a fox-tail to a long antenna giving us enough time to take evasive action. We both came to a sudden halt at the crest, radiator to radiator. We paused for a friendly discussion about birds of prey and birds in general (avian species, not *Homo sapiens*) then went our respective ways.

From Cameron's Corner it is only sixteen kilometres to 'Bollard's Lagoon' Station homestead where George and Elaine Riek always insisted on a 'cuppa' or a meal and a bed if we had time to stop. A hundred kilometres further on, at 'Merty Merty', Ted and Pam Riek were just as hospitable. Ted and George are twin brothers and between them they have six and a half thousand square kilometres of country. Since they are outside the fence the area is dingo-infested and unsuitable for sheep, so Ted and George run cattle only. The brothers had been earth-moving contractors around the Northern Territory for many years before settling in South Australia. Their station area was an isolated one for many years until the discovery of gas and oil at nearby Moomba. Ted was a superb raconteur, surpassed only by the head stockman Ross Scobie, and we spent many pleasant hours listening to their anecdotes about life in the Outback. Ross was an ex-station owner and had grown up with the Aboriginals, speaking their Dieri language fluently. He claimed that his grandmother, who had travelled by camel from the rail-head at Gawler, was among the first white women the Cooper Aboriginals had seen. The

Cresting one of the numerous sandhills along the dog fence on the Queensland-New South Wales border.

Scobie name became synonymous with the Birdsville Track where many descendants ran several stations. Ross' grandfather, apart from being a pioneer settler, also operated the mail service along the Track.

We arrived at 'Merty Merty' in the evening and spent the next day hunting for raptor nests - and what a day that was! We found Black Falcons, Brown Falcons, Little Falcons, Little Eagles, Wedge-tailed Eagles, Whistling Kites and Black-breasted Buzzards. We'd found one of the latter while it was still being built in August and it now held two eggs. Two other Buzzard nests were empty, although they were well-lined with green eucalypt leaves ready for egg-laying. We were slightly apprehensive our inspection would lead the birds to desert them. This proved short-lived since they continued to occupy both nests and a further inspection on 6 October revealed two eggs in one of them. Two days later the other nest was still empty.

On 7 November we were amazed and mystified to find the nest held two tiny chicks - one showing the start of wing and tail primaries and at least a week old, the other a day or two old and very tiny. We pondered the smallness of the chicks, but the apparently short incubation period puzzled us most. We believed the eggs of a raptor that size would take six weeks to hatch, not the three or less indicated here. We wondered if they had been incubated elsewhere and then transferred to the nest, but we dismissed that notion because the Buzzards had been occupying the nest at every visit we'd made since finding it on 17 September.

We took a couple of close-up shots of the chicks in the nest and gathered some bird legs from it which appeared to be Kestrel legs. Then we decided to try to film the chicks being fed since there was only one chick in the other nest we'd been working near the 'Merty Merty' homestead. The tower and hide were quickly set up and we retired discreetly to observe the adults' reaction to it before attempting any photography. The Buzzards circled the nest-tree for the next two hours and as the sun grew hotter we began to fear for the chicks' lives. Finally we removed the tower and retired to the homestead nest. The following day we returned to the area and, finding both adults on the nest we assumed the chicks were alright and didn't disturb them.

On 26 November, accompanied by Albert Chamberlain, we returned to the Strzelecki and found the chick in the homestead nest had disappeared. There was little else for us to do there, so we decided to press on and work an Australian Hobby nest near Innamincka.

The next day we headed south again to work the Buzzards with the two tiny chicks - now estimated to be half-way through their fledging period. Driving through the heat, we saw up ahead a light truck parked off to the side of the track, beside it stood a young woman. We thought her vehicle had broken down and Albert, a latter-day Sir

Walter Raleigh, held his coat at the ready - although a parasol would have been far more appropriate. She stood bareheaded and lovely, poised and cool, in the blazing sun. We pulled up beside her, but before we could speak she said, 'Dr. Schodde is just over there with my husband shooting Chestnut-crowned Babblers for museum specimens'. Albert's face fell. Despite the heat he would have enjoyed helping her.

She was Jenny Weatherly. We'd last seen her at the Australian Broadcasting Commission in Melbourne where, as Jenny Hobson, she worked with Mike Vance on the Wild Life Australia series. She was travelling with her husband, Richard, and Dr. Richard (Dick) Schodde of the Wild Life Division of the C.S.I.R.O. in Canberra. It once again goes to prove what a small world we live in.

Dick Schodde and Richard joined us and told how they had just come down from the Birdsville area in Queensland. Apparently they had passed by while we were working the Hobby nest sited a few hundred metres off the track. They confided they had found Black-breasted Buzzards nest-building there and offered to reveal the location. We appreciated that, but doubt if we were at first believed when we told them we had five Buzzard nests in this area from which to choose. As we were on our way to work one, they accepted our offer to accompany us to it.

The Buzzards' nest with two tiny chicks and the remains of other nestling birds.

As we approached the nest tree a Buzzard left the nest hurriedly. At the foot of the tree crouched a young kestrel, obviously fallen or knocked from its nest somewhere. It was fully fledged but still had some down clinging to its feathers. As Dick Schodde stooped to pick it up, it flew, obviously for the first time, to the cover of a fallen tree branch close by. At the same time another Kestrel perched close to the Buzzard's nest flew down to the ground near to us and put on a 'broken wing act', no doubt in an effort to entice us away from the one on the ground. Richard Weatherly followed it in an attempt to get a photograph of this unusual action. He returned shortly without having got a shot as it evaded him in the grass. He commented that he wasn't aware that Kestrels did that 'act'. None of us had ever seen a Kestrel do it before.

We talked for a while at the foot of the nest tree while the Buzzards circled overhead voicing alarm calls, and then the trio left for their homes. The tower was set up at 17:00 and we watched from a distance to see the Buzzard's reaction. They accepted it readily and were back on the nest brooding before dark, so preparations were made to begin filming next morning.

At 07:00 the tower was moved in to six metres from the nest and the hide entered. Instead of two half grown Buzzard chicks the nest contained four chicks of varying ages; they were all Kestrels. A few moments later three more flew into the nest and mixed readily with the four younger ones. One was showing some down and was undoubtedly the one that Schodde had attempted to catch at the foot of the tree the previous afternoon. The other two flying ones could have been the two tiny chicks we'd photographed in the nest on 7 November, as this tied in with the known fledging period of three to four weeks for the species.

It is difficult to express in words the photographer's first reaction to this bizarre situation and for some time he was convinced he was dreaming and would awaken eventually to prove it so. All manner of explanations raced through his mind and one after the other they were reasoned out and rejected. He was one who strongly believed every happening could be explained by cause and effect. Nothing came by chance, there were no ghosts, Holy or otherwise, yet still he shivered and momentarily came out in goose-flesh in the hot atmosphere of the hide. He wondered if they were the victims of an elaborate hoax, but that didn't hold up either when all facets of the situation had been studied.

The shots of the two chicks taken on 7 November were still in the 35mm camera which we used for close-up shots of eggs or chicks in the nest. When the film was eventually processed it was easy to see they were indeed Kestrels. People may wonder why we, who had worked both species before, failed to recognise them in the first place. We were expecting the Buzzards to be rearing their own kind and even though they were much smaller than expected we thought perhaps that could have been through the birds breeding for the first time and possibly laying very small eggs. This behaviour was unprecedented anywhere in the world so there was nothing to arouse our suspicions in this case. The mistake - if one could call it that - was acceptance of the norm by association.

The flying juveniles fluttered their wings and

Erecting the tower at the Buzzards' nest.

called expectantly for food, as one of the Buzzards circled the nest-tree voicing alarm calls. Then it flew in and alighted on the nest with a sprig of eucalypt leaves. It stayed momentarily, no doubt alarmed by the noise of the Beaulieu, and flew off to return with a fledgling bird which appeared to be a Kestrel - their usual prey at this nest. No attempt was made to feed the chicks clustered about soliciting to be fed, although one of the flying juveniles fed itself from the fledgling still held in the Buzzard's talons. The cameras' noise was distracting the bird and it continued to voice alarm calls. It was also being vigorously attacked by a Magpie-lark, which had a nest of almost fully-fledged young higher in the tree. The Buzzard eventually flew off leaving the carcass of the fledgling Kestrel with the chicks.

The Buzzards soon settled down to feeding their large brood normally. The attacks of the Magpie-lark continued, however, giving the Buzzards no peace unless they left the nest; the Magpie-larks never bothered the Kestrels which, of course, posed no threat. It is difficult to explain why the Magpie-larks should attack the Buzzards so relentlessly. The Buzzards presented no threat to the Magpie-larks in the vicinity of the Buzzards' nest.

Like most raptors they always hunted further afield and took Magpie-larks as prey on some occasions. The Magpie-larks seemed aware of their immunity or they would not have risked nesting in a tree occupied by Buzzards. Until we can learn another reason for their aggressive behaviour we'll just have to put it down to downright pugnacious cussedness.

A Buzzard was about to brood the chicks when she suddenly flew off. The seeing-out party had arrived as arranged. Five hours had never passed so swiftly. A somewhat perplexed photographer was trying to restrain his excitement about the possibilities seemingly emerging. If it wasn't a dream or a hoax then it must be one of the strangest aberrations of animal behaviour ever recorded.

Since it was extremely hot we drove to the Mundibarcooloo waterhole to cool off for the afternoon, leaving the cameras in the hide while we were gone. When we returned they were almost too hot to touch. The late afternoon sun was creating some possible flare problems as well, but they were overcome by improvising a larger lens-hood.

The next morning we made another early start and recorded some interesting footage of the chicks being fed and brooded. We collected some

The Buzzard brings green leaves to the nest which now has four Kestrel chicks of varying ages.

more bird-leg debris and later sent the samples to the Wild Life Division of the C.S.I.R.O. in Canberra. Our earlier suspicions were subsequently confirmed: they were Kestrel legs.

A thunderstorm appeared to be building up so we decided to move out at 09:30. While we were lowering the tower, the Buzzard, which had taken up its usual station circling the nest-tree, was attacked by a Raven which was in turn attacked by the Buzzard. The Raven, apparently realising it had stirred up something more troublesome than it could safely handle, took off, pursued by the Buzzard. After chasing the Raven out of sight it returned and swept through a nearby Coolabah, emerging with a Kestrel, grasped by a wing, in its talons. The Kestrel, struggling and chattering, was carried a considerable distance before it broke free or was released by the Buzzard. A very shaken Kestrel made it back to the Coolabah, where it just managed to retain its perch. The Buzzard did not strike again. We're certain the Kestrel was one of her foster chicks, and was only attacked because the Buzzard was upset by our activities or the presence of the raven.

It has been suggested an egg collector had exchanged the Buzzard's eggs for those of Kestrels to keep the Buzzards in occupancy of the nest. If there had only been those first two Kestrel chicks in the nest that would have been a plausible explanation, but to account for seven of them someone would have had to collect another five of varying ages to place in the nest along with those two we had photographed on 7 November. We found no evidence that anyone other than ourselves had been near the nest. The area at that

time was a rather remote one and the nearest track seldom used as it had been impassable for much of that period due to heavy rain.

As we were worried as to the possibility of our film being heat affected, we thought it prudent to make another trip a week later. We left home at our usual 03:30 starting time and had the tower in position by 18:00. The Buzzards were becoming relatively tame, as one would alight on the nest while we stood at the foot of the nest tree six metres below. The prey continued to be fledglings, but they were no longer Kestrels. This wasn't surprising as they must have taken all the Kestrels available in their zone of operations.

We stored the tower at 'Merty Merty' station for use in the following season. After a very hot trip, with two punctures and collapsed front end on the ute, we arrived home at 01:00 Tuesday 7 December, having travelled two thousand kilometres, as well as doing our work at the nest, in forty five and a half hours.

Some rather bizarre behaviour had been observed and photographed. This pair of Buzzards were robbing Kestrel nests, bringing the chicks to their own nest, then rearing them on other Kestrel chicks and, possibly adults. The number of legs in and around the nest indicated their diet in the earlier part of this episode, was largely Kestrel. We are unable to account for this behaviour, but we are positive it was not due to human interference.

One possibility we offer is that the female was being fed on the nest by the male while she was endeavouring to lay, and some of the chicks, being alive when brought in, have given their open-bill

The female Buzzard feeds the four remaining Kestrel chicks.

soliciting for food. The stimulus presented by that open bill and the soliciting sounds usually accompanying it has caused the Buzzard to respond by feeding it. We have on several occasions seen live prey brought to the nest but in each case it was eventually eaten. An instance of other than the parent birds feeding chicks is described in a chapter on the Black Falcon.

Another possibility, (a very remote one, we think) is that a frustrated female, unable to produce eggs, has gone and deliberately kidnapped the chicks.

When we link this behaviour with that of this species breaking Emu eggs with a stone to get at the contents we may well need to re-assess the parameters of the bird brain - especially this species. Although we regard the rearing of those chicks to be aberrant behaviour the idea remains inconclusive as little or no study of the species has previously been done at the nest. We feel certain this wasn't the first time this pair reared other than Buzzard chicks. In a previous year Jack Purnell, of Sydney, found a nest in the same area (possibly the same tree) as this pair, that had four chicks. We've asked Jack, since this episode, if he was sure they were Buzzard chicks. His reply, 'Of course they were Buzzard chicks, the Buzzard was on the nest'! We respect Jack's knowledge of birds generally, but we aren't convinced he was right on this occasion. The chicks could have been Black Kites, which are hard to distinguish from Buzzard chicks without close examination and a thorough knowledge of the fledglings of both species. We have inspected a lot of Buzzard nests and we never found more than two eggs. In

most cases only one chick was reared.

Another puzzling aspect of this pair's behaviour was the rearing of seven chicks when one or two was the normal brood. The fledging duration of those seven Kestrels was spread over six weeks: this is the fledging period for one or two Buzzard chicks. Was the three to four week Kestrel fledging period too brief for the female to lose her mothering instinct?

An interesting sidelight was the action of the flying juveniles in putting on the 'broken wing act' to protect the younger ones still confined to the nest. At our last visit this was demonstrated again while we were getting close-up shots of the four not yet flying. Two of the flying ones came around the nest-tree and then flew to the ground to demonstrate other versions of being in trouble. One clung to the base of a small Coolabah and flapped its wings around it and chattered loudly, while the other chattered and fluttered its wings seemingly unable to fly or caught by its feet on a low branch. These tactics are not uncommon among some avian species and are employed to draw an intruder's attention away from the nest or young, but are usually performed by the parent birds. Here we had young birds attempting to protect other young with which they had no kinship except that they were all reared by the same foster parents. The protective role adopted by those juveniles did not extend to feeding time as they competed with the others for the food brought in. Luckily there was plenty for all.

In the following year the Buzzards prepared their nest and kept it freshly lined with eucalypt leaves for several weeks. It was as far as they

69

went. They would have had difficulty in finding suitable foster chicks in the prevailing conditions. We observed several pairs that year and all failed to breed.

Some of our footage of this episode was televised in August 1978 and a few viewers doubted its authenticity. Perhaps this description will help to convince them it was not a hoax perpetrated by us - or by others on us. With a little more of the same luck that put us in the right place at the right time to record this behaviour, we should be able to obtain further information on this intriguing pair, if they ever give a repeat performance.

In April 1979, a pair of Buzzards, which we assumed to be the same pair, occupied the nest for a while, lightly lining it with green leaves. Visits in early and late May found them nearby but the nest had not been further lined. Late in June there was no sign of the Buzzards near the nest, but a few kilometres to the north we found three Buzzards near a newly constructed nest.

At first we assumed the third bird to be a juvenile from the previous season, but a closer scrutiny showed it to be a much older bird. On our next visit in mid July only two Buzzards were seen in the vicinity, one of which alighted on the nest but only stayed briefly. We assumed the third bird had been driven away with the approach of the breeding season, but returning three weeks later, we again found three birds present, with the nest now fully lined.

A further three weeks ensued before we again visited the nest to find one bird in it and a second perched in the tree above. There was no sign of the third bird. The nest was still empty.

About one week later, on 7 September, a Buzzard, which we assumed to be the female, left the nest as we approached. From one hundred metres away we watched the male bring a small rabbit to the nest, where he was soon joined by the female. After taking possession of the prey, she drove off the male with two aggressive lunges towards him, with wings spread. She fed for a few minutes, then stood quietly in the nest for about fifteen minutes, when the male returned and copulated with her. He remained on the nest for five minutes before leaving, voluntarily this time.

Within a few minutes he was back, and copulation again took place. Leaving the nest, he flew to the ground about seventy five metres away, picked up a stick and returned to the nest, where he again copulated. The two birds then spent some time arranging sticks in the nest, before the male again left. He had flown less than a hundred metres when another Buzzard landed on the nest, and, after a minute of inaction by both birds, copulation took place for perhaps twenty seconds. He left the nest shortly afterwards to join the first male, soaring high in wide circles about a kilometre away.

Unfortunately, being busy elsewhere, and not realising the significance of this behaviour, our further observations were limited to three brief visits. The first, three weeks later, found the female incubating two eggs. After a further three weeks she was still incubating, but on our final visit in November the nest contained one rotten egg and a week-old chick, which indicated the eggs were laid about ten days after the two males were seen to copulate with the female. On each of the last three visits only one or two Buzzards were seen at the nest, but the visits were brief with no effort to look for the third bird.

Although we were aware that polygamy - one male to two or more females - was common in some raptor species, we were unaware of the relative rarity of confirmed polyandry - two or more males per female.

Apparently this behaviour has only been recorded in two raptor species - Harris' Hawk, *Parabuteo unicinctus* (Mader 1975) and the Galapogos Hawk, *Buteo galapogoensis*, (DeVries 1975). In these species, polyandric groups were sometimes quite common, with the male birds (up to three of them in the Galapogos Hawk) all copulating with the female and participating in incubation and food provision. [10] Although we did not see the second male Black-breasted Buzzard participating in nest helping, it seems highly likely that he did.

Mader suggested that polyandry in Harris' Hawk in Arizona may be an adaptation that increases nesting success in a desert environment where food resources are scattered and subject to fluctuations. DeVries noted that the Galapogos Hawk nests in the drier, more barren areas of the Galapogos Is. and that polyandric groups were more frequent in dense hawk populations. We found it interesting that these conditions related to the Black-breasted Buzzards in 1979.

However it is perhaps more significant that the only documented instances of polyandry occurred in *Buteo* and the closely allied *Parabuteo* genera. There has been much discussion about the origins of the Black-breasted Buzzard and its relationship to other species, the most popular theory today being that it has evolved from Kite-like stock and is in fact an aberrant Kite, despite its Buzzard-like appearance. We feel however, that if, as is indicated here, polyandry does occur in this genus, then it could perhaps, indicate stronger links with the *Buteo* genus.

The Buzzard watches over her brood of Kestrel chicks.

SPOTTED HARRIER
Circus assimilis

When viewed from close quarters this species is probably Australia's most beautiful raptor. We've watched this Harrier hunting in an apparently leisurely fashion low over cereal crops, stubble or open grassland. Along a fence-line, with a slight crosswind, they work in a series of arcs, taking a few wing-beats to windward of the fence, then soaring low along it and drifting to leeward, they repeat the few wing-beats to windward again. This move is so precise one is drawn to counting the wing-beats to see if they vary in number. They seldom have.

The Harrier favours plains and open wooded grassland and is found around cultivated areas throughout Australia. It is generally common in the Inland, although rarer near coastal districts. It is nomadic, its movements associated with the availability of food. In some years it is very common in the Victorian Mallee, whereas in other years it is absent. We saw it in most of the Inland areas we travelled, and worked the species both in the Mallee and as far away as the Birdsville Track.

Their prey consists of small mammals and birds. In nests we've worked we saw rabbits, rats, mice and birds fed to the chicks. The birds were often Quail and Pipits since these are generally ground dwellers and are the most likely species to fall victim to the Harrier's hunting methods in our region.

In 1974 we worked a nest in the Mallee where

A typical clutch of Spotted Harrier eggs.

the pair of Harriers took domestic chickens. This didn't endear them to the farmer's wife. She asked us to let her know when we'd finished so she could get her revenge with a shotgun. We pointed out their advantages around the farm, such as keeping down the number of mice and rabbits, and suggested she pen the chickens for a few weeks. She agreed to think about it.

Green leaves were brought in several times a day to reline the nest. They were generally eucalypt, though occasionally some other species. Both birds usually shared this work. At some other nests we sometimes found the male reluctant to come to the nest with leaves or food. This could possibly have been due to the proximity of the hide inhibiting him.

The Spotted Harrier in flight.

A female Spotted Harrier about to feed the dominant chick of her brood.

The Harrier chicks have a peculiar habit of standing in the nest with heads bowed in an obsequious manner. At first we thought it a response to a warning call from the parents. However, it would continue after an adult alighted on the nest with prey and we became convinced it was a result of a pecking order established among the brood. Standing thus they were not so likely to be attacked by the most dominant one. It was very noticeable at feeding time, although it actually started when they were anticipating the arrival of prey. Should the domineered ones have the audacity to stand upright before the topdog had been fed, or had a substantial amount, he would peck them back into place - their submissive stance. When there came a lull or loss of tempo in the feeding rate of number one the next in line would appear to sense it and would take a peep around to see if number one was sated. If it was, the movement would be ignored and the second in line would be fed. If the prey was substantial, a rabbit for instance, all would eventually be fed in turn, but if mice or small birds, the ones down the line

The female Harrier shades her chicks.

73

The use of two towers allowed some choice in positioning the photographer at the Harriers' nest.

could get very hungry and become more aggressive, thus cancelling out the dominance of the one normally above, and communal feeding might take place in relative harmony. At some nests the aggression of the dominant one has had us contemplating mayhem. Occasionally we would work a nest where they all got along together like a well disciplined family of *Homo sapiens*. This seemed to happen where the adults behaved well toward the photographer, their acceptance of him being exemplified in the behaviour of the chicks toward each other.

In 1977 we worked a nest of three chicks in a Sandalwood Tree, only fifty kilometres from home and had high hopes of working them from early fledging to flying, but a severe hail and wind storm on 4 October blew them from the nest, killing them. The same storm did millions of dollars of damage in the Sunraysia District of north-west Victoria.

In October 1978 there was an influx of the species into the Millewa farming area where they had a distinct preference for nesting in trees bordering the ripening cereal crops. Trees bordering fallow, stubble or open grassland had little

The female Harrier alights on her nest in the top of a Mallee Tree.

appeal as nesting trees that season. Hunting seemed confined to the cropped land or the fence-lines bordering it. The nesting trees were either Mallee, or Belar. The relatively large rabbit population was probably the main contributing factor to the large influx of the species. They, like many of the other raptors were about one month later breeding than they were in the previous season.

We set up at our first nest to be worked in 1978 on 4 November. The two chicks had hatched the previous day. A single seven metre section of tower was exactly the height required but we found that some days it was not positioned right in relation to the wind direction to get the best landing shots, so we set up a second tower which gave us some latitude of choice in positioning the photographer. It proved an ideal solution as it led to a superb landing shot. The female was very wary and the male seldom came to the nest at all while we were working it. We left one tower **in situ** and concentrated on another nest a few kilometres away where both adults behaved well and shared the hunting, feeding and supply of green leaves to the nest. It was also one of those nests where the chicks behaved well toward each other.

On 9 December 1978 we returned to our first nest, entering the hide before sunrise. One chick had been killed falling from the nest. The other chick, now thirty-nine days old, took little notice of us, although we knew it was capable of flight if we approached too closely. About sunrise it began calling for food, then suddenly flew from the nest. This appeared to be its first flight since it made a clumsy landing in the top of a leafy mallee tree. It called for half an hour, then flew back to the nest-tree to make yet another clumsy landing in the foliage and scrambled through it to the nest.

At the second nest the birds continued to behave well. When we found one chick missing at only twenty-eight days old, we wondered if it had fallen from the nest and been eaten by a fox or feral cat. However, a couple of days later we found it back in the nest and it flew as we approached; apparently it must have flown voluntarily two days earlier. The second chick did not leave until twelve days later: a fortnight's difference in fledging.

At another nest a female Harrier arrives with leaves.

The Harrier chicks crouch low as the female arrives with a piece of rabbit.

The pair of Harriers shared nest duties. The male is noticeably smaller, but darker plumaged.

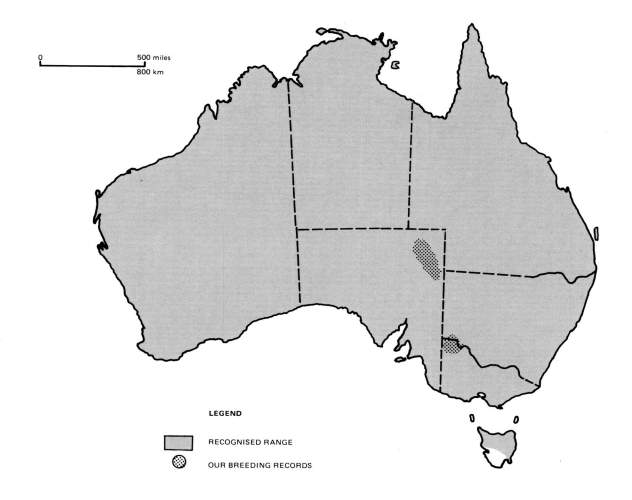

SPOTTED HARRIER   *Circus assimilis*

circus- hawk (Gk); *assimilis*- similar (L).

OTHER NAMES: Allied harrier; Jardine's harrier; Spotted swamp-hawk.

LENGTH: 510 - 600mm. Female larger than the male.

WINGSPAN: Approximately 1200 - 1300mm.

DISTRIBUTION: Common in many inland areas throughout Australia, near plains and wooded grasslands and around cultivated areas. Appears to be nomadic and possibly migratory. Rare in coastal districts - Also found in islands to the north-west, from Timor to Celebes.

VOICE: Alarm call is a rapid **'kit-kit-kit-kit'**; the food call is a loud whistling **'seep'**; at other times, **'kitter-kitter-kitter'**, as communication between the adults.

PREY: Small mammals - particularly rabbits and, during inland irruptions, rats; and birds - particularly quail and pipits.

NEST: A large flat structure lined with green leaves, usually placed relatively low in a tree and often well-hidden amongst foliage. It is 500 - 600mm in diameter and 200 - 300mm deep. We have found nests as low as two metres above ground level, but more often at heights from 4 - 8 metres. On two occasions nests were at a height of 14 and 15 metres respectively.

EGGS: Two to four, usually three, 53 x 41mm. They are coarse, dull bluish-white ovals, unmarked, but often stained. We recorded 21 clutches: 7 with four eggs, 13 with three, and 1 with two. Egg-laying is usually from August to October. However, while most of our records show egg-laying in September and October, we have found eggs in April, July and August.

MARSH HARRIER
Circus aeruginosus

This species is sometimes known as the Swamp Harrier, Allied Harrier, Gould's Harrier or Swamp Hawk. The Marsh Harrier, as it is officially known, and the Spotted Harrier are of a similar size, but the difference in plumage makes them easily distinguishable - the Marsh Harrier having a white rump patch clearly recognizeable in flight as it banks and turns over the open grasslands, crops, lakes, coastal mudflats and swamps which comprise its usual habitat. It is common in the north, east, south-east and south-west of Australia and rare in the arid inland, although during 1974 and 1975 we often saw them hunting on Sturt's Stony Desert. We believe they are permanent residents around the bores, only extending into the desert in wet years.

The Marsh Harrier builds a rough nest of weeds and very light sticks, usually in swamps and reed beds, although it has been known to build a nest in a standing cereal crop. Each nest that we worked was in reeds growing in water of a depth of one to one and a half metres, with the base actually in the water. Three or four eggs formed the normal clutch.

This species was the most difficult to work suc-

cessfully. We found them exceptionally wary with apparently acute hearing and a reputedly keen sense of smell - a possible reason why they seemed to avoid a hide near a nest, although we feel that after wading through an odorous swamp to reach a nest it would be difficult for the bird to distinguish between the photographer and the surrounding malodour. Care must be taken when setting up the hide and visiting it or the birds may desert the nest. Even searching for the nest and

Bringing camouflage for the hide.

The Harriers' view of the hide.

78

A typical clutch of Marsh Harrier eggs.

Downy Harrier chicks, which usually have distinctive white "skull-caps".

coming upon it suddenly while the eggs are being brooded can be enough to cause desertion.

The position of the nest in the reeds can be fixed by watching over the swamp. If eggs are being brooded the male will be flying in with prey for the female which is usually transferred in the air - the female flying below him and deftly catching the prey as he releases it. When the chicks are a week or two old both birds will be bringing prey to the nest. Searching can be time-consuming. In large swamps with hectares of tall reeds we've taken days fruitlessly searching for a nest we were sure was there. Once enveloped by the reeds all landmarks disappeared so it was difficult to move in a straight line. We solved this particular problem by taking a compass-bearing on the point where the birds were seen to drop into or emerge from the reeds, then following it. Invariably we found the nest.

Hide preparations must be carried out with both caution and speed. After studying the breeding times from previous seasons, searching can be timed to begin after hatching or late in the incubation period when the birds have, hopefully, become more attached to the nest and less likely to desert it. The chicks need to be at least a week old before the hide can be prepared, and the weather moderate because the small chicks need constant brooding in extremes of temperature. Preparations must be quick. We never spend more than half an hour near a nest in any one day. The hide must be down-wind from the nest and well-camouflaged. It is wise not to work the birds too closely, especially at first. Usually we set up about ten metres from the nest, then as our technique improved, moved in as close as four metres.

The female Harrier brings dry grass to line the nest.

79

The female Harrier alights with a frog.

We had one desertion to blot our copybook. We began preparing the hide soon after the three chicks hatched and had it in and camouflaged in a quarter of an hour. However, returning two hours later we saw the adults still circling high above the nest. We dashed in to remove the hide. We were too late. Two chicks had died and the third was nearly dead. We had blundered badly and knew that it was from inexperience and a certain amount of impatience. It taught us not to prepare the hide at this most critical period, nor do some initial work then drive off hoping for the best. We learned to wait and observe from a discreet distance to see if and when the hide was accepted.

After a few days the adults would settle down and we would bring in the cameras. This is one bird we suspect may be able to count a little so wherever possible we have a seeing-in party of more than one person - and a seeing-out party if we plan to return. So that the birds don't have to become re-accustomed to objects on the hide when the cameras and flash-heads are replaced we leave bottles and shiny tins in their stead. These tactics are important and have become routine. Nevertheless, we have had some long waits. At one nest we sat a total of twenty-nine hours over a six-day period before getting an adult to the nest. After four or five hours in the hide each day we would leave to allow the chicks to be fed. Then, toward the end of their fledging period, the female suddenly settled down and devoted her time to feeding the chicks, which for weeks had been feeding themselves from prey brought in by her and dropped into the nest. For the first time we were able to observe an adult Marsh Harrier behaving normally. She fed the juveniles, carried away scraps and would often return with her bill full of weeds which she spread on the nest.

At another nest we photographed three chicks about a fortnight old. When we returned the following day we found one chick pecked to death. During the day the other two chicks ate it, the legs being the last to go. There is little doubt the killing and cannibalism was caused by the proximity of the hide. The absence of the adults causes stress among the chicks. As they become hungry their aggression increases.

A variety of prey was fed to them - including water birds, rabbits and lizards. There was quite often some amusing confrontations between the chicks for the prey brought in, as at some nests the adults would not stay to feed them. They would stand erect facing each other with the disputed food lying between them. One would slowly move its head down toward the food, at the same time intently watching the other as if expecting a kick in the head; a stomp of a foot was enough to bring that head up very smartly and the confrontation would continue. There was a lot of feinting with the feet by both birds before one would lead with a left or right foot thrust to the body which was generally countered simultaneously by one from

The Marsh Harrier in flight.

A Harrier chick bathes in the water near its nest.

A fully fledged Harrier chick.

Female Marsh Harrier.

Male Marsh Harrier.

the other. The wings were held high and wide and slapped down on each other in occasional rallies. The bill was not used and there appeared to be no damage sustained by either bird in these encounters. We were fortunate in capturing this action of cine film.

At six weeks the juveniles were fully fledged, and they were much darker than their parents, the breast feathers being almost black.

On 25 November 1977 we found a nest containing three chicks about ten days old - they were still a downy off-white with no sign of emerging wing or tail primaries. The nest was in a small clump of isolated reeds which meant we had to bring reeds from another area and transplant them where we intended to place the hide. The job took several days because we only stayed a maximum of half an hour near the nest in any one day. We were ready to begin photography on the 30th.

The male bird often brought prey to the nest, usually rabbits carried in his talons, but he never stayed to feed the chicks. Possibly he never feeds the chicks although we can't be sure the nearness of the hide wasn't a deciding factor in his short stay. The female was also wary but her concern for her chicks overcame her fears long enough for her to feed them. She soon learned to ignore the flash and the camera shutter. The whirr of the Beaulieu was another matter and we decided to stop filming an obviously frightened bird. She seemed to prefer frogs and water-birds as prey, whereas the male brought in rabbits. Sometimes she would meet him as he approached. Once he was carrying a rabbit and appeared to be making heavy going with it. As he released it she flew beneath him and with split-second timing caught it and flew to the edge of the swamp where she fed from it. Later she flew in with a portion to feed the chicks.

On hot days the chicks drank from the swamp and occasionally waded on the submerged edge of the nest. They were always alert to the return of the parent, and their eager, but subdued calls became the signal to switch on the power for the flashheads in time to capture landing shots.

Attempting to garden close to the nest or take close-up shots of the chicks always triggered an aggressive reaction from them. They would stand threateningly, hiss and sometimes drop onto their backs and strike out with their talons.

By Christmas Day they were fully fledged and flying short distances to nearby reed clumps. They returned to the nest to be fed. By now the hide was a mere 3.8 metres away and both adults were still bringing in weeds and sticks, filling the centre cup depression, and building a rough platform. However, the adults never entirely lost their wariness. The male stayed only long enough at any one time for one shot to be taken, and the female continued to feed the juveniles before flying away with any scraps. We still had some long waits between visits by the adults.

On 28 December high winds and squally conditions kept rabbits in their burrows and birds were the only prey, some of them being Australian Crakes. Approaching the hide two days later we found two of the juveniles perched on the top cover. While heavily laden with camera gear, standing knee-deep in the black odorous ooze and water up to our waists, we managed to take a few final shots of them there. Our deliberate caution with the preparation of the hide and the careful movement of it toward the nest had paid off handsomely. The chicks were now over six weeks old. It was time for us to move on.

The dark plumage of the Harrier chicks contrasts with that of the female.

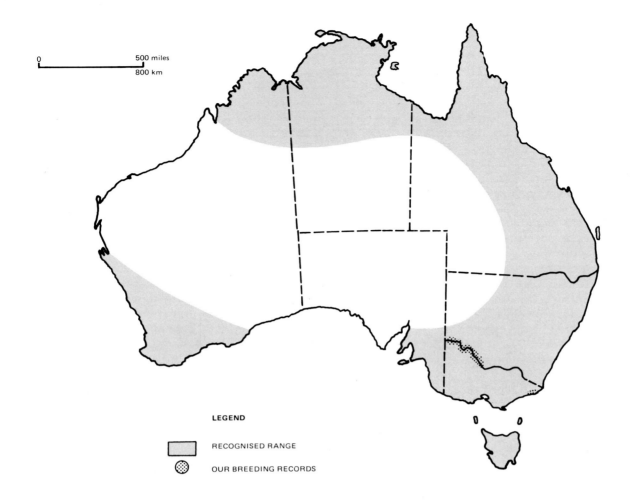

LEGEND

▭ RECOGNISED RANGE

⊛ OUR BREEDING RECORDS

MARSH HARRIER *circus aeruginosus*

circus- hawk (Gk); *aerugo*- rust of copper (L).

OTHER NAMES: Swamp harrier; Allied harrier; Gould's harrier; Swamp-hawk.

LENGTH: 500 - 580mm. Female larger than the male.

WINGSPAN: 1200 - 1300mm.

DISTRIBUTION: Common in the north, east, south-east and south-west of Australia; in open grasslands and crops near swamps, lakes and coastal mudflats. Very rare in arid inland areas, but will appear in high-rainfall years. Birds in the south and in Tasmania are usually migratory while elsewhere they are generally sedentary. The Australian subspecies *C. a. gouldi* extends to south-east New Guinea, New Zealand and south-west Pacific Is. Other subspecies range over much of Europe, Asia and Africa.

VOICE: In courtship a short, **'kee-a'** or **'kee-o'**. The food call is a high-pitched whistling, **'psee-uh'**.

PREY: Birds, small mammals, frogs, reptiles and large insects. Carrion is also eaten. At nests we worked rabbits were the main prey, but birds (particularly water birds), frogs and lizards were recorded.

NEST: A large platform of very light sticks, weeds and reeds, lined with grass, usually placed in swamps or reed beds where it may be in water or on the ground. Occasionally a nest may be built in a standing cereal crop or long grass. All nests we found were amongst reeds growing in water one to one and a half metres deep. A typical nest measured 600mm in diameter and was 300mm deep.

EGGS: Three to five, 50 x 38mm, rounded ovals, dull bluish-white, unmarked. We found four nests with eggs: two with three eggs and two with four. Eggs are laid from August to December in the north, October to November in the south. Our records show egg-laying from early to late October. Eggs hatch in 33 days and chicks fledge in about 42 days.

BLACK FALCON
Falco subniger

This species is thinly scattered over much of inland Australia. It inhabits open and lightly timbered country.

As with other falcons they do not build a nest of their own. They usually occupy the deserted nest of either another raptor or a corvid species. In the Strzelecki area the old nests of the Black-breasted Buzzards are sometimes used. A typical clutch consists of three eggs, although two to four are often found. The eggs are rounded ovals and coloured pinkish-buff heavily spotted with red-brown.

In the inland feral cats play havoc with nestling birds.

A clutch of Black Falcon eggs.

We had seldom seen Black Falcons prior to 1975 when we discovered a nest containing four eggs in a beefwood on 'Clifton Hills' Station in Sturt's Stony Desert. The Outside Track of the Birdsville Track passed quite close to the nest-tree and the incubating bird was flushed by each passing vehicle. Fortunately there was little traffic at the time to have a detrimental effect on incubation, but the nest could be clearly seen from the road, making it an easy target for egg-collectors. We were pessimistic about the eggs remaining undisturbed. They were photographed in the nest.

Following our return home we mentioned to one or two people that we'd worked Letter-winged Kites on 'Clifton Hills' Station. A local identity told us he'd never seen this species before and he decided to observe them in the colony we had worked. We made a special point of revealing the exact location of the Falcon nest, hoping it would forestall the great temptation those eggs represented either to him or to anyone else in his party if they were to come across that conspicuous nest. He assured us that he 'wouldn't dream of taking the eggs under the circumstances'.

When the party returned we asked them about the eggs and whether they had hatched. We were told there was nothing in the nest and that feral cats had probably taken them. That old saying, 'truth will out' held good as one of the party unintentionally let this particular cat out of the bag - and it wasn't a feral one either: one of them had indeed taken the Black Falcons' eggs. We had wasted our time, our breath and our trust. The eggs are now in a local collection.

Despite our disappointment we returned to the area accompanied by David Hollands. On 16 July 1975 we found another nest. Later we found two more a few kilometres from the first.

We approached this species with the utmost caution as we believed they would be difficult to work. This seemed to be the case when we saw the female leave the nest every time we turned off the Track a kilometre from her. We therefore set up the tower well back from the nest and moved it closer in stages. Surprisingly they proved relatively easy to photograph. The female quickly returned to the nest and both birds continued their hunting, feeding and brooding, apparently oblivious to us.

The Black Falcon in flight.

Setting up the tower at a Black Falcons' nest on Sturt's Stony Desert.

The female Black Falcon with three downy white chicks.

The downy-white chicks were brooded constantly between feedings. The female left the nest briefly to collect prey from the male perched in a nearby tree. He usually followed her back to the nest-tree, perched higher up and kept a close eye on the feeding while he carefully preened himself.

The male always hunted well away from the nest area despite the abundance of prey nearby. From the hide he could be seen disappearing across the gibber, probably to another watercourse a few kilometres to the south. Birds were plentiful along all the watercourses and they formed the bulk of the prey at the nest. We noticed Yellow-throated Miners, Magpie-larks and on one occasion an Inland Dotterel being fed to the chicks. Rats were brought in occasionally. The rats are generally considered nocturnal, but it proved there was some movement by some during daylight.

The female's plumage was very grubby, in contrast to that of the male. Her constant brooding of the chicks meant there was little time to spare for preening. When day-brooding ceased, however, she was able to devote much of her time to herself. She soon presented herself as immaculately as her mate. Often she would move out to a limb closer to the cameras and perch totally engrossed as she carefully nibbled each feather from base to tip, oblivious to the sounds of the cameras from the hide a few metres away.

Satisfied with our coverage of the chicks at this stage of their development we pulled out and took the Track south, but not before debating whether we should have headed north-east to Birdsville and then south as the sky to the south had looked ominous for the past few days. David couldn't comprehend what the worrying was about. His trips into the inland prior to this one had been by aircraft and thus he had no idea how quickly, nor to what extent the tracks could deteriorate even under relatively light rainfall. We made excellent progress till we reached the Cooper where some rain had fallen. From there on the Track became progressively worse as we ran into more rain on a track that had obviously already received some over the past few days. About ninety kilometres from Marree we realised it was hopeless, the Track was a quagmire and David was at the point of collapse from his efforts to keep the vehicle mobile by pushing on many occasions. Much of the night was spent intermittently dozing while sitting up in the front of the vehicle. Sometime during the early hours of the morning it was realised that David no longer occupied the passenger seat and one got a reasonably good nap.

Day brought a most dismal scene. It could be likened to the moors of England on a winter morning, a bleak, water-laden, treeless plain swept by mist and rain; and out of the rain appeared David carrying a sodden sleeping bag. Apparently unable to sleep while sitting up he had in desperation, taken his sleeping-bag and slept, so he said, on the Track. Some humans are like trooper's horses, able to sleep while standing,

Stuck on the Birdsville Track.

while others apparently need to be horizontal.

There was little we could do except sit and wait till the weather improved. The annual average rainfall for the area is very low but the rains of 1973 and '74 were fresh in our minds when many had been stranded and had to be fed by airdrops, while others had been lifted out by helicopter. We had plenty of food, and water wasn't a problem, except that at the moment there was far too much of it. David was, however, finding it hard to accept the situation as he had planned to be back at his surgery in Orbost, two thousand kilometres to the south, on the morrow.

The rain stopped about mid morning so we started walking as we knew Dulkaninna Station homestead was only a few kilometres to the south. It was extremely heavy going as the clay soil becomes very sticky after the surface water drains away. Several kilograms of clay had to be removed from our boots after a few struggling steps as they became too heavy to lift, or as in one case, being elastic sided, tended to get left behind. Bare feet would have been better had ones hide been tough enough to withstand the sharp edges of broken gibber mixed with the clay. We made it to the homestead, but probably because it was a Sunday, radio contact could not be established with the outside world so we trudged back to our vehicle.

The sun was shining, which coupled with a strong wind was improving the Track surface by the hour. A hot drink and a meal made everything look brighter and almost anything possible, so an attempt was made to drive on. Once mobile we kept going at a pace that carried us through the worst spots. From Marree we were able to phone our wives, but told them we were not optimistic about getting home for a day or two going by the glimpse we had of that south-bound road as we drove off the Track into Marree. It looked as if a squadron of tanks had been holding manoeuvres on it. With more than a modicum of luck, coupled with the expertise acquired through a lifetime of driving on outback tracks we managed to keep mobile most of the time, although we weren't always facing in the homeward direction. The occasional sound of a sharply indrawn breath indicated the occupant of the passenger seat was not asleep. David was learning that a couple of thousand kilometres in inland Australia could be other than a few hours in an aircraft. On more than one occasion he was heard to mutter, "its unbelievable".

We arrived home in time for breakfast.

The next trip was a month later, and as it was planned to drive on to Birdsville and thence across Queensland, Pamela elected to be co-driver and seeing in party. She had been most disappointed the previous year when we'd found our quarry, Letter-winged Kites, south of Birdsville and so didn't get to see that outpost of the 'Never Never Land'.

The chicks, which had been snow white on our previous visit, were now darker than their parents. After a few shots for the record and a little cine we moved on toward Birdsville. Soon after leaving the falcon's nest we were at the point where we'd seen our first Letter-wings the previous year. Had we not been lucky enough to spot that lone bird then, we now found we would have had to travel right to Birdsville before finding them, as there were no more trees for one hundred kilometres. We were in the longest paddock we'd yet encountered, being at least one hundred and sixty kilometres from gate to gate. In it were some of the ten thousand head of cattle usually run on Clifton Hills station.

Soon after crossing the northern boundary of South Australia we were in Birdsville. Its not the smallest township in Australia but close to it.

Once a year it becomes a Mecca for thousands of race-goers and others. The others probably going to see how the others cope. They come from all over the Commonwealth, by all manner of transport from aeroplanes to motor cycles. There are few amenities but plenty of beer, wine and spirits. Empty cans become more than ankle deep inside and outside the pub. This littering is encouraged apparently to give the place an 'atmosphere'. Personal opinions about that atmosphere varied widely. But the races were over and we saw only four people in the few minutes it took to top up our fuel tank. During that brief stay Pamela managed to photograph the town and half its population. Those two, holding up the pub verandah post were apparently camera shy but not quick enough to evade her camera as they moved inside.

We moved on to look for the Square-tailed Kite on the other side of Queensland and then home. A six thousand kilometre trip.

As we approached the Black Falcon's nest on our next trip, one adult flew low over the nest tree. The young had flown but one had been caught by the leg under a sliver of wood on the trunk of the tree where a limb had broken away. It had only just died and probably accounted for the adult's presence as we approached. It was disappointing to miss out by such a small margin of time in getting a record of their final days in the nest. Fledging would have been between six and seven weeks.

The female Falcon alights with a Native Rat for her half fledged chicks.

While checking nests along the Strzelecki Creek on 16 August 1976 we found one with two eggs, which from their size and colouring we assumed to be those of Black Falcons. Thirty two days later we checked it again and found three eggs, one of which was about to hatch as it was cracked and movement of the chick within could be heard. As we still hadn't been able to catch a glimpse of a bird leaving the nest, we scanned the sky with binoculars and were lucky enough to spot both falcons circling at great height above the nest tree. Assuming the first egg was laid on 14 August, incubation would be thirty four days. We were intrigued to find the nest well lined with river washed sand that was not there on our first visit. It was too coarse and in too large a quantity to have come from the feathers of the adults or be blown there. We believe it was put there by the birds themselves and it could have only been carried in the bill. We worked them when the chicks were about a week old and left the tower *in situ* when we went home.

On returning in October we found the tower down and two chicks dead on the ground. Reports from nearby Moomba gas fields indicated winds of up to one hundred and sixty kilometres an hour had torn a path through the area. The single section of tower was not damaged and we resumed working. Rabbits were the only prey we saw fed to the remaining chick.

A tour of the area in April 1977 found several of the species still about. It would appear that they are to some extent sedentary, probably while prey remains reasonably available. With conditions deteriorating through an almost total lack of rain

A female Black Falcon with her brood in an abandoned Kites' nest.

in that area during 1977, many Black Falcons moved out to more favourable areas to the south to breed in the Spring. We saw several around our home where the species is considered rare. A pair nested and reared three chicks there.

Two years later, in 1979, we were delighted to be shown a pair nesting nearby. They were using an abandoned corvid's nest, placed six metres high in a Sandalwood Tree. When first inspected,

Rabbits formed the bulk of the prey at this Black Falcons' nest on the Strzelecki Creek.

it contained one newly hatched chick and three eggs. A few days later there were three chicks and one egg, but when we erected the tower we found the brood reduced to just two chicks. They were about two weeks old and beginning to fledge.

Our presence near the nest provoked repeated vigorous attacks from both adults. They soared high overhead, then stooped steeply to within a few metres of us. However, once the hide was entered the female soon settled down to feeding the chicks on such birds as Quail, Pipits and Galahs.

Unfortunately she was a dull and untidy bird, and as her more dapper mate was not coming to the nest, we decided to concentrate on other species. We made our last visit to the nest when the chicks were five weeks old and nearly fully fledged.

At our first Victorian nest, the female Falcon feeds a bird to her eighteen day old chicks.

Between July and November 1979 we found more Black than Brown Falcons along the Strzelecki. While searching for the Grey Falcon we found and checked twelve nests of Blacks, most of which had three chicks. Apart from those twelve we probably missed as many as they usually see the searcher first and slip quietly away at low altitude till well clear of the nest. They may then on occasions be spotted circling wide at great height.

David's tower was set up at a nest with three small chicks in October 1979. The adults attacked vigorously as the tree was climbed and the nest checked. David and two other enthusiasts worked on it. One, Ray Martin witnessed a very unusual episode in bird behaviour. The Black Falcon chicks at one stage were obviously hungry and quite vocal about it, but the adults were absent, possibly hunting. Their open bill soliciting for food must have been too much for an Australian Kestrel nesting about fifty metres away as she came in and fed them. Ray obtained several shots of the episode.

In October 1980 we set up on a nest in our home

Two flying juvenile Black Falcons try unsuccessfully to feed on a bird.

The dark plumage of the juvenile Black Falcon contrasts with that of the parent bird.

district that had four almost fully fledged chicks, two of which made their first flight as we approached. The third appeared capable of flying but remained in the nest with the smallest which obviously couldn't fly. The adult birds attacked vigorously as the tower was climbed but no one was struck. The two flying young were soon back on the nest awaiting to be fed. The male brought in a small passerine which each chick in turn attempted to feed from, while the adult looked on.

It was obvious they had never fed themselves before, which makes them dependant on the adults for some time after fledging. The male flew off as the female arrived and she fed them in turn. There was no jostling or fighting among them as there is among many other species. Their table manners were exemplary. The adults' plumage was a lighter brown than any of the species we'd worked before and contrasted strongly with the almost black plumage of their young.

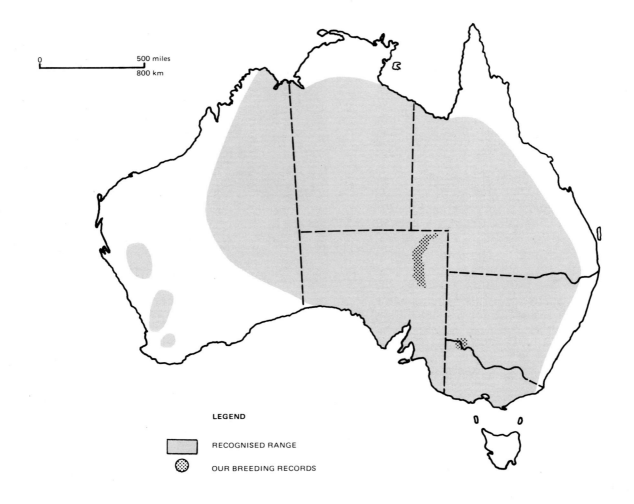

LEGEND

RECOGNISED RANGE

OUR BREEDING RECORDS

BLACK FALCON *Falco subniger*

falco - falcon (L); *subniger*- somewhat black (L).

LENGTH: 450 - 550mm. Female larger than male.

WINGSPAN: Approximately 950mm.

DISTRIBUTION: Generally rather rare. Frequents open and lightly timbered regions of the inland and sub-interior of the Australian mainland. Nomadic and somewhat irruptive; may become locally numerous. Not found outside Australia.

VOICE: Deep **'gak-gak-gak-gak'** or a drawn-out moaning **'karrrrr'**, both calls being given near the nest.

PREY: Birds, particularly ground birds, mammals and insects. We recorded native rats, rabbits and birds such as Miners, Magpie-larks, Dotterels, Pipits, Galahs and Starlings.

NEST: The old nest of another raptor or corvid species is generally used. We found the nests of Ravens, Black Kites and Black-breasted Buzzards were most often used.

EGGS: Two to four, usually three 54 x 40mm. They are rounded ovals of pinkish buff, heavily spotted with red-brown. Of thirteen clutches we found, four were of four eggs and nine had three. Eggs were laid from June to December, our records showing egg-laying from June September. The incubation period is 34 days and the chicks fledge in 42.

BROWN FALCON
Falco berigora

A dark phase Brown Falcon shades her chicks.

A light phase Brown Falcon at her nest.

The difference between the light phase and dark phase of this species has, in the past, caused it to be recorded as two different species. In 1912 A. J. North wrote: 'I therefore intend to keep the two separate although I may not be correct in doing so'. There is certainly a large variation in colour, some of it depicted here.

The species is sometimes known as a Brown Hawk and at other times Cackling Hawk because of its loud, coarse cackle. The scientific name derives from the Aboriginal name for the bird *berigora*.

The Brown Falcon is common in a wide variety of habitats throughout the mainland and Tasmania. It is absent only from dense forest. There are five recognised subspecies in Australia, and it is also found in southern New Guinea.

On rare occasions the species has been recorded building its own nest, but it usually claims the deserted nest of another raptor or corvid. It sometimes nests in an open tree hollow.

We worked the dark phase of this species along the Birdsville Track in 1974. There were four young in the nest and a fifth dead on the ground beneath it. It was not surprising that the fifth was unable to retain its place in the nest as it was overcrowded even with four. Their prey was entirely native rats during the short time we worked the

The Brown Falcon appears at its best in flight.

nest. At that time we were content to take only a few photographs of a species then move on to the next.

In 1976 we worked the light phase along the Strzelecki Creek. The chicks hatched just after the hide was entered. The female appeared excited - she kept up a low chatter as she stood on the edge of the nest peering into it. As each chick hatched she carried the egg-shell from the nest and then brooded the chicks constantly. She did not call for food at all that day. She seemed content to sit and brood.

Next morning the hide was entered early in an attempt to film the chicks being fed for the first time. It was to be a long wait. The female sat and made no call for food until early afternoon. She called for a while and when there was no response she flew out to where the male was perched in a tree about fifty metres away. She chattered loudly for a few moments and returned to the nest.

After a short brooding session she repeated the entire peformance: a few calls from the nest, a short wait for a response from him, then the short flight to him. This action was repeated many times over the next two hours. Then he suddenly flew off. In a few minutes he was back at his perch and she joined him almost simultaneously to collect a tiny lizard or skink. It was a meagre meal when divided between two small, twenty-four hour old chicks, but it seemed to suffice. The female must have been hungry too, having eaten nothing for the past twenty-four hours, but she contentedly brooded her chicks. A very tired, cramped and hungry photographer descended the tower seven and a half hours after ascending it.

In November 1977 we found a nest of three eggs near our homes. Two eggs had hatched on 17 December and the third the following day. The youngest chick was three days old when we entered the hide, placed only 3.8 metres from the

The female Falcon arrives with a skink for her six day old chicks.

The eggs of the Brown Falcon.

nest. Within minutes of the seeing-in party leaving, the female was back on the nest brooding. The male hunted and soon called the female off to collect a small skink. One chick swallowed this whole. A few minutes later the male again called the female off the nest and she collected a small rabbit. This more than satisfied the chicks. The chicks constantly appeared ravenous and while being fed they often pecked the female about the head and bill in their eagerness. On one occasion a chick got hold of her tongue and held on. It was intriguing to see the length and elasticity of it and her reaction when the chick's grip was finally broken. She shook her head several times, stared at the chicks for a few moments, and went on feeding them.

We noticed the female carried mice, small birds and spiders in her bill, but anything larger such as rabbits were carried in her talons. The male never came to the nest, although some reputedly do so occasionally. The female always carried any scraps from the nest after feeding. Even so there were many ants at the nest and she spent considerable time pecking at and removing them. Ants occasionally get the upper hand and can cause a chick to become demented and fall from the nest.

On 2 January 1978 the hide was entered without the usual seeing-in party. The wait for the female was much longer than usual, not unexpected under the circumstances. She brought in a small bird which she quickly dropped into the nest and departed. One chick managed to half-swallow it whole, but was then unable to move it either further down or disgorge it. It lay flat in the nest for a while, then attempted to clear it again. Once more the chick was unsuccessful. Its distress calls

seemed to be heard because the female quickly returned. She grasped a protruding leg and pulled the offending body out. Then, appearing to forget the presence of the hide occupant she went on feeding her chicks normally. Several times we witnessed such an automatic response to such an emergency, though none as critical as in this case. The chick would certainly have died without help.

When the hide was entered ten days later one of the four week old chicks was dead. We did not know the cause of its death. After the female fed the remaining chicks she flew off with the dead one. By 30 January the other two chicks had left the nest, although one was still unable to fly and was being fed on the ground. There had been a violent rain-squall a few days previously, so it was possible it had been blown from the nest. Fledging was thus between six and seven weeks at this nest.

The Brown Falcon chicks at five weeks.

The female Falcon feeds her fourteen day old chicks.

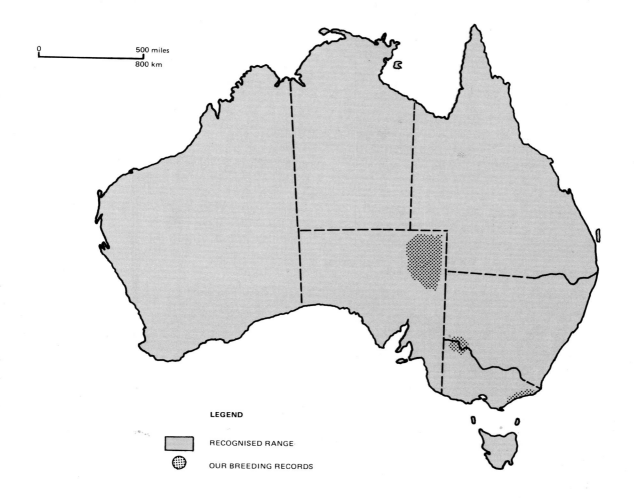

LEGEND

▨ RECOGNISED RANGE

⦿ OUR BREEDING RECORDS

BROWN FALCON *Falco berigora*

falco -falcon (L); *berigora*- Aboriginal name for this bird.

OTHER NAMES: Brown hawk; Cackling hawk.

LENGTH: 430 - 500mm. Female larger than the male.

WINGSPAN: 900 - 1000mm.

DISTRIBUTION: Common in a wide range of habitats throughout Australia; absent only from dense forest. Nomadic in the inland, elsewhere sometimes sedentary. Five recognized subspecies in Australia: *F. b. berigora*- coastal and highland districts of east and north-east Australia. *F. b. tasmanica*- Tasmania and Bass Strait islands. *F. b. centralia*- mainly the arid interior. *F. b. occidentalis*- south-west Australia. *F. b. melvillensis*-- coastal north Australia and offshore islands. Also southern New Guinea.

VOICE: A loud **'kee-arr'** or a loud, somewhat hoarse cackle.

PREY: Small mammals, birds, reptiles and insects. We recorded native rats, mice, rabbits, small passerines, crickets and spiders.

NEST: Usually the deserted nest of another raptor or corvid, but on occasions an open tree hollow. This falcon is also reported to build its own nest on rare occasions.

EGGS: Two to five, usually three 50 x 38mm. They are rounded ovals, fine but without gloss, pale buff, spotted and blotched to varying degrees with red-brown. We recorded twelve clutches: 3 of four eggs and 9 of three. We also found a nest containing five chicks. Eggs are usually laid from June to November; our records indicate egg-laying from late July to mid-November. The incubation period is not less than thirty days and the chicks fledge in about forty-five days.

PEREGRINE FALCON
Falco Peregrinus

Races of this species are found throughout the world. In Australia it is known for its spectacular stoops and hunting methods. Though generally sedentary it will become nomadic under certain conditions. The Latin *peregrinus* in fact means 'wandering'.

There are two subspecies in Australia: *F. p. macropus* in most of the mainland and Tasmania, and *F. p. submelanogenys* found only in the south-west of Western Australia. It inhabits coastal and inland cliffs and gorges, timbered watercourses and lightly timbered country. It sometimes appears in urban areas.

It nests in a variety of sites. The most common are on rock ledges or cavities in cliff faces where eggs may be laid in a scrape in the earth. Stick nests are also used on rock ledges or, in the interior of Australia where cliffs are less common, a Wedge-tailed Eagle's nest. The nests of smaller raptors or corvids and the hollows of trees are sometimes used. The Peregrine often occupies these eyries all year round and will defend them

A female Peregrine brings a Galah to feed her three chicks.

98

even outside the breeding season.

The Peregrine usually collects its prey in flight, plunging with extreme accuracy at great speed. There are claims that one has been clocked at 275 mph (442 kph) [11]. In one plunge near Bendigo, Victoria, in May 1977, a Peregrine hit the flashing blue light of a police car. The light was smashed and the bird was killed. It was the second such attack in the area in two years. It would appear that they have no more respect for the police than they have toward photographers.

In the 1960's the Peregrine became the subject of considerable concern in the world because of its rapid decline in many areas. Studies found that the decline in populations had begun in the early 1950's.

The decline coincided geographically and contemporaneously with the introduction of chlorinated hydrocarbon insecticides. The best-known of these was DDT.

Further studies, for example those by Derek Ratcliffe in Britain and J. J. Hickey and D. W. Anderson in North America, have documented the effect of insecticides, especially DDT, on the Peregrine and some other birds of prey. When DDT enters the food chain it becomes concentrated as it moves higher. Birds of prey are particularly vulnerable to DDT contamination, Peregrines, Ospreys and Bald Eagles feed off prey already contaminated and suffer magnified contamination. Concentrated insecticide residues in birds cause thin egg-shells and the resulting high incidence of breakages has led to the decline of the Peregrine.

Investigations by the Victorian Fisheries and Wildlife Division have tended to confirm these

The Peregrine in flight.

A typical clutch of Peregrine eggs in a nest hollow.

overseas findings. The Peregrine is not a common bird, but more eyries are being found than were previously known. We have seen more of this species in Victoria than at any previous time, a hopeful sign of their survival.

A female Peregrine at her nest hollow.

The female Peregrine in an impressive stance.

In January 1975 we found an eyrie not far from home and kept it under discreet surveillance during the summer and winter months. It was an old Wedge-tailed Eagle's nest at a height of twelve metres in a belar. Any approach closer than about a hundred metres brought loud protests from both birds.

On 11 October 1975 we were sure they had chicks so we set up the tower well back from the nest-tree and allowed them a few days to accept it. While erecting it we were under constant attack from both birds, but we weren't actually struck. The speed with which they attacked and the closeness of their approach left no room for complacency. We wasted no time at the nest site because apart from the threat of being struck we didn't want to run the risk of desertion or harm to the chicks through unnecessary exposure to the heat. They were estimated to be about a week old.

We began filming five days later. The seeing-in party of four had hardly left when the female returned to the nest. The adult birds were initially distracted by the noise of filming, but they soon settled down to feeding and brooding.

On our second visit to the nest the female left the chicks exposed to direct sunlight while the photographer was climbing the tower and setting up cameras and flash-heads. The chicks became heat-distressed. As soon as the seeing-in party moved off the female returned and stood between the chicks with outstretched wings. With a quivering motion she fanned the chicks for several minutes until they were quiet and apparently dozing in the shade. It was a wonderful display of parental concern.

The chicks were fed approximately every two hours. Both adults attended the nest and shared the feeding to some degree, although the female did all the brooding and about two-thirds of the feeding. Birds were the only prey brought in, most commonly Galahs. Although the adults were acting naturally during filming they always attacked us, voicing their alarm call - a shrill and rapid **kak-kak-kak-kak,** when we approached or departed the hide. The female was the most aggressive, coming within centimetres of our heads, unless we turned to face her. The male, although vocal, seldom came closer than a couple of metres of us.

A violent storm in early November blew the tower down, but fortunately left the chicks unharmed. The tower was only slightly damaged and was re-erected on 10 November when we photographed the fully-fledged chicks. It proved good timing as they left the nest the next day. They had been forty-two days fledging. We were unable to establish the incubation period precisely because we are reluctant to disturb them when they are about to lay - but our estimation is between thirty and thirty-five days.

In 1977 we discovered more nesting pairs than we had ever seen before, all in old Wedge-tailed Eagles' nests. We worked one with two chicks. However, a hail-storm on 4 October left one chick dead. We were surprised to find one had survived.

The following year the same nest contained three eggs, although only one chick was reared. In 1979 there were again three eggs, all of which hatched, and we worked these until the chicks were fully-fledged.

A Willie Wagtail harasses the female Peregrine.

The female Peregrine feeds her three tiny chicks.

The fully fledged Peregrine chicks.

A young Peregrine prepares for its first flight.

In 1980 the three eggs in the nest once again produced only one chick, a very poor performance on average. On hot days the female repeatedly used her bill to try and push the chick's head into the shade of her body, giving us some nice photographs. The chick moved around the nest with no apparent sense of danger and when it was just beginning to show signs of emerging wing and tail primaries we found it on the ground near the nest-tree. Mysteriously it managed to survive the fall of almost ten metres and was, in fact, quite unharmed. We returned it to the nest. It learned nothing from its fall. It continued to approach precariously close to the edge of the nest and a fortnight later it was on the ground again. This time it did not survive.

The female Peregrine feeds her single chick in an old Wedge-tailed Eagles' nest.

The female Peregrine tries to shade her chick.

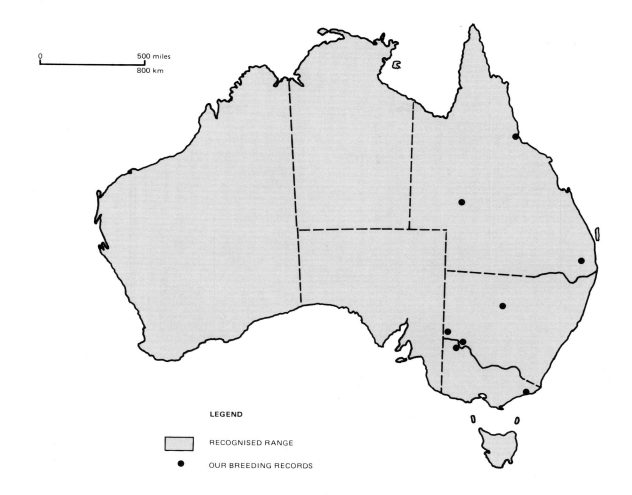

LEGEND

| | RECOGNISED RANGE |
| ● | OUR BREEDING RECORDS |

PEREGRINE FALCON   *Falco peregrinus*

falco -falcon (L); *peregrinus*- wandering (L).

OTHER NAMES: Black-cheeked falcon; Duck hawk; Pigeon hawk.

LENGTH: 380 - 480mm. Female much larger than the male.

WINGSPAN: Approximately 950mm.

DISTRIBUTION: Generally uncommon, throughout Australia, near coastal and inland cliffs and gorges, along timbered watercourses and in lightly timbered regions; sometimes in urban areas. Mostly sedentary in Australia. Two Australian subspecies: *F. p. macropus*- most of Australia including Tasmania. *F. p. submelanogenys* - South-west Australia. Other subspecies range over most of the world.

VOICE: Alarm call is a loud 'kak-kak-kak-kak' or 'kek-kek-kek', becoming more rapid and shrill during attacks. Also near the nest a drawn out keening 'kreeee-ee'.

PREY: Chiefly birds, but small mammals, frogs and fish have been recorded. At nests we worked only birds were recorded as prey. One pair in north-west Victoria appeared to feed largely on Galahs, along with a few Starlings and other passerines; while in south-east Victoria prey was largely Parrots, particularly Crimson Rosellas. Also identified as prey were Domestic Pigeons.

NEST: A variety of nest sites are used, the most common being a rock ledge on, or a cavity in, a cliff face where eggs are laid in a scrape in the earth. We also saw used a stick nest on a rock ledge. Throughout the interior, where cliffs are less common, the nests of Wedge-tailed Eagles are favoured, but sometimes that of a smaller raptor or corvid. Open hollows are also used.

EGGS: Usually two or three, rarely four, 51 x 41mm. They are rounded ovals, creamy buff, heavily marked with red-brown and dark-brown spots and blotches. Of nine clutches recorded by us: 7 were of three eggs, 1 of two eggs and 1 of four eggs. They are laid from August to November, our records showing egg-laying from early August to mid-September. Incubation is about 30 days and fledging about 42 days.

103

AUSTRALIAN HOBBY
Falco longipennis

The Australian Hobby is also known as the Duck Hawk, Black-faced Hawk or White-fronted Falcon, but is best-known as the Little Falcon. It is Australia's second smallest falcon, being 300 - 350mm in length, with a wingspan of 660 - 800mm. These dimensions would appear to rank it with the Australian Kestrel as the smallest falcon, but the Hobby is much the heavier of the two.

At first glance the Hobby appears to be like a small Peregrine, though closer scrutiny shows it to have more rufous and less barred underparts, and a less extensive black cap - having a pale forehead and partial collar. Like the Peregrine it is a very courageous hunter and also defends its nest vigorously.

Male Australian Hobby.

Two fully fledged Hobby chicks on Marrapinna Station, New South Wales; the third chick was a week later fledging.

Although generally not very common, the Hobby is found all over Australia. Though favouring open wooded areas it also frequents a variety of habitats and is often seen around city parks and gardens. The few breeding birds we have found have been in drier inland areas where they usually select old corvid's nests, these being generally placed near the top of tall trees. They show a decided preference for those along watercourses, but this may be because they are taller there.

In October 1974 we found a nest with three chicks in a tall gumtree on the Noonthorangee Creek on 'Marrapinna' Station one hundred and sixty kilometres north of Broken Hill, New South Wales. It was beyond the scope of our tower at that time so we lowered the nest in stages, to about eighteen metres from where we could work on it. To keep the nest intact during this manoeuvre we wrapped it in wire netting. At each staging the nest was left long enough for the chicks to be fed at least once before attempting a lower position. The strategy was quite successful and we obtained some good shots.

The chicks were fed on other small birds caught by the male and transferred to the female in a nearby tree. They were usually mutilated beyond recognition when she brought them to the nest. The chicks were by this stage feeding themselves and one would grab the prey as she alighted and turning quickly away from the others, mantle over

it. After a few unsuccessful attempts by them to get it they would accept the situation and just watch the other feed. While not feeding or watching a nest-mate feeding they spent a lot of time preening and exercising their wings. A small rise in wind velocity was nearly always the signal for a session of wing flapping. One of the chicks appeared to be about a week behind the other two in development, but it was not due to missing out at mealtime, at least not while we had it under observation. We'd noticed at several raptor's nests that there could be up to a week between the laying of the first and last egg in a clutch.

The Australian Hobby in flight.

On 14 October 1976, along the Strzelecki Creek, we set up on a nest with three eggs. We'd found the nest occupied on 17 September but had refrained from inspecting it then, as inspecting a nest before eggs are laid could possibly lead to desertion. They proved to be a delightful pair to work, taking little notice of flash or camera noise even though we were less than four metres from them. The tower was left *in situ* when we returned home. Four weeks later we found two chicks and a rotten egg in the nest. We estimated the chicks to be about a week old, so eggs could not have been laid till late September, vindicating our caution in not inspecting the nest in mid-September.

A White-breasted Woodswallow was one bird brought in as prey that could be identified. There was a lively tussle between the chicks for this bird, and honours were about even as the prey alternated between them and was quickly disappearing. Only the wings remained with most of the feathers intact. With a wing each and facing the cine camera, it looked like an organized race to see which would finish first. The lead alternated, and with only a few centimetres of wing protruding, they rested a while; eyeing each other, they started again, and the feathers disappeared into their respective bills simultaneously as the cine ran out of film. A fitting result, and one that has raised a cheer on occasions where it has been shown at service clubs.

While working at this nest, which was near a water hole, we often watched dingoes come in to drink and sometimes bathe. Some were beautiful specimens of the species, and no doubt pure dingo, while others were obviously crossbreeds by various domestic dogs. Many were comparatively tame, and often very curious, as they would circle our vehicle on occasions when we pulled up near

A typical clutch of Hobby eggs.

them. We once came on an old-man kangaroo bailed-up, and we were able to witness first-hand the legendary cunning of the dingo. The dog stood facing the 'roo, while the bitch lay resting a couple of metres behind him. As we pulled up beside the trio, the 'roo hopped slowly away. The dogs followed till one of them moved past him and barred his way. Their strategy was quite evident, they weren't allowing him to eat or drink until he was too weak to defend himself, and then they would move in for the kill. They had a decided advantage over him in so far as it would only take one of them to keep him from eating, while they would no doubt hunt and eat other small game, while waiting for the eventual feast. While we were frantically trying to get the cine camera unpacked, the dogs' attention was diverted for a few moments and the 'roo moved across the Track close to the front of our vehicle, putting us between him and his attackers. Apparently believing he had a chance to escape his executioners, his loping hop changed abruptly to a nice turn of speed. The dogs hesitated for a moment, then

A dingo comes to drink at a waterhole near the Hobbys' nest.

skirting past us took off after the old-man. He had no hope against their speed, and before he'd travelled a kilometre they had him bailed up once again. We would have liked to have followed and filmed the drama to its inevitable end, but we had neither the time nor the off-road vehicle necessary for the task. We drove on leaving the old-man to his fate, for what right had we to interfere in the natural order of things? This drama, (to us) had after-all been going on for thousands of years.

Besides dingoes, thousands of Little Corellas came in to drink both morning and evening. They were a most beautiful sight, but a nuisance and a potential hazard. A nuisance because their noisy gatherings made it impossible to sound-record the Hobbies at the nest, and a hazard because they clustered on the guy-ropes of the tower to swing by their bills. It was impossible to do any filming while the tower swung alarmingly, so a hole was cut in the side of the hide and the Corellas filmed as they did 'catherine wheels' while hanging by their bills. There was no way of knowing how much damage was being done to those ropes, but they held till the cine ran out of film, and of course we're pleased we took the risk to get such enchanting behaviour recorded on film. It has proved to be as enjoyable to the viewers as it obviously was to those performing birds.

While the female Hobby was brooding eggs or small chicks, we didn't have to wait long for her return to the nest. When day-brooding finished, however, her visits were few and far between.

The chicks were banded when about a month old. One fluttered from the nest while we were engaged in the task but we were able to catch and return it. Fledging was estimated to be about five weeks.

The female Hobby feeds her two chicks in an abandoned Kites' nest.

The female Hobby with her partly fledged chicks.

In an abandoned raven's nest, a female Hobby feeds her three tiny chicks.

107

At three weeks the Hobby chicks need little help to dispose of a Parrot.

In October 1979 we found and worked another nest on the Strzelecki. The chicks were just hatching, and the female remained on the nest while the necessary gardening was carried out beside her. The tower and hide was set up 1.8 metres from her. We could have worked closer had we desired, but being too close tends to inhibit natural behaviour. There were three chicks but one of them only survived for a day. We filmed the surviving chicks being fed a Zebra finch.

A visit a couple of weeks later coincided with her being off the nest to collect another finch from the male. She dropped the plucked finch as the tower was being climbed, and attacked continuously from almost any direction, weaving her way between tree branches and guy-ropes, to graze the intruder almost every time. As he entered the hide she struck his pack containing camera gear. Feathers fluttered down but she appeared unhurt, although when inspecting the results of our photography some weeks later, we noticed a bare patch of skin near her left shoulder, where she had apparently struck something.

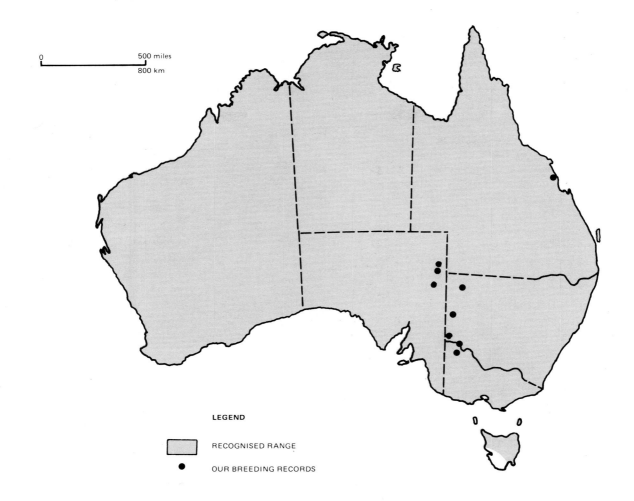

LEGEND

▭ RECOGNISED RANGE

● OUR BREEDING RECORDS

AUSTRALIAN HOBBY   *Falco longipennis*

falco - falcon (L); *longus* - long (L); *penna* - feather (L).

OTHER NAMES: Little falcon; White-fronted falcon; Duck hawk; Black-faced hawk.

LENGTH: 300 - 350mm. Female larger than the male.

WINGSPAN: 660 - 800mm.

DISTRIBUTION: Generally uncommon; throughout Australia, particularly in open wooded country, cultivated areas and in city parks and gardens. Probably generally sedentary.   Two Australian subspecies: *F. l. longipennis*- Southern Australia including Tasmania. *F. l. murchisonianus*- Interior and northern Australia. Also southern New Guinea and neighbouring islands.

VOICE: Shrill, rapid **'kee-kee-kee'**.

PREY: Mainly small birds and insects.   We recorded only birds: those identified include Zebra finches, Red-backed parrots and Wood swallows.

NEST: Usually the stick nest of another species. Nests we found were usually deserted corvid's nests in the topmost branches of a tree. There have been reports of the species building its own nest, but this is considered doubtful.

EGGS: Usually three, occasionally two, rarely four, they measure 46 x 34mm. They are rounded ovals, pale buff and glossless, heavily freckled with brown and chocolate. We found five clutches all consisting of three eggs. Eggs are usually laid from September to November. However, on one occasion eggs were laid in early January (This was the second brood for the season). We were unable to establish the incubation period, but fledging appears to be about 35 days.

GREY FALCON
Falco hypoleucus

This falcon is one of Australia's rarest raptors, generally found only in the drier inland of Australia along timbered watercourses. It is nomadic, although probably nests in the same area each year. We found it nesting along the Strzelecki Creek and sighted it several times in that same area. We had one sighting of the Falcon outside this general area. This was at Pirlta, in the Millewa district of Victoria.

Our first sighting was in July 1974 when a pair had taken possession of a large Mallee tree with a large stick nest in the topmost foliage. We could approach quite close to them before they would fly to a nearby tree. When approached again, they would fly back to the nest-tree. A photograph of one was taken to confirm identification. We'd only started the project a month earlier and were elated, as it appeared that we would be getting the near-rarest raptor first. However, approaching the area a few days later we found cars belonging to a pipeline crew parked under the nest-tree and there was no sign of the Falcons.

The next sighting was on the Birdsville Track when a pair flew noisily into a Coolabah near our camp, one evening in July 1975. David must have identified them from their calls as his excited, 'Grey Falcons!' was uttered before they appeared in the evening gloom. They were gone when we awoke next morning, indicating their movements are not confined to daylight hours.

We were on our way down from the Birdsville area in August 1976 and decided to have a look along the Strzelecki Track and Creek. It'd been a long day and we'd long since run out of conversation, and for lack of something more sensible to say, the driver said, 'I'd be happy if we could find

Crossing the Cooper at Innamincka.

a pair of Greys before dark. It doesn't take much to make me happy, you know'. 'Of course not', came the reply with more than a hint of sarcasm. Less than a minute later the vehicle was stopped and binoculars called for. There was a pair of grey birds, partly obscured by foliage, perched in a gumtree at least a hundred metres off to the right, on Pelican Creek. They were Grey Falcons! We walked over to take a shot of them for the record. The extraordinary coincidence of the conversation, coupled with the luck of a one-eyed driver spotting that pair of Falcons, did not lead on to the working of them. They were kept under observation periodically for some weeks before they disappeared. After such an auspicious sighting we would have wagered on them nesting there. It tends to reinforce that proverb 'Don't count your chickens before they're hatched'.

In November 1977 we saw a pair near Duarlingie

The desert after rain.

A pair of Grey Falcons on Pelican Creek.

Waterhole on the Strzelecki Creek, but we were unable to find a nest - although we searched a wide area.

We saw a pair on the Coongie Track a few kilometres north of Innamincka in May 1978 while travelling with David and his family. They disappeared in the direction of the north-west branch of the Cooper that fills the Coongie Lakes on occasions. We went on to the lakes which were rich in bird life generally, and among the more notable species saw Bustards and Brolgas. One moonlit night, we were fortunate enough to see and hear the Barking Owl near our camp. We returned in August to hunt for the Grey Falcons but without success. We did find Black Falcons nesting, and plenty of evidence of the aborigines' earlier occupation of the area in the form of artifacts and work-sites.

Our sixth sighting came toward the end of our fourth trip into the inland in 1979. We'd made two trips along the Birdsville and Strzelecki Tracks, and two northward from 'Merty Merty' along the Strzelecki and part of the Cooper Creek - about twelve thousand kilometres in all. Many other species had been found, which tended to break the monotony and frustration. We'd done more walking of watercourses than usual on the fourth trip and were footsore, weary and somewhat disheartened, when we stopped to check the nest of a 'tame' Wedge-tailed Eagle we'd worked in 1976. There was a bird on the nest, so we walked over to see if it was the same one. It was. As we stood near the nest-tree marvelling at her calm acceptance of us, a pair of Grey Falcons flew low overhead and one alighted on a nest only about a hundred metres away. The Wedge-tail was forgotten. We watched for awhile, a little uncertain as to whether we should inspect the nest, but decided against doing so and drove on to 'Merty Merty'. We returned next day, but found no sign of the falcons. A Little Crow occupied the otherwise empty nest and thus we knew the falcons were still nest-hunting. It was a let-down after the euphoria of the previous day, but we were optimistic that we'd find them nesting close by in a week or two.

A fortnight later, on 4 August, we returned to what we thought would be a very short search. However, it was not till late on the third day that we sighted a pair, twenty-eight kilometres north as the crow flies from where we'd seen them on 20 July, though more than a hundred kilometres along the route taken by us, as we combed the Strzelecki flood-plain. There was no proof that they were the same pair, but they probably were, as we'd made a very thorough search of the whole area. One bird was perched on a stick nest in the topmost branches of a slender Coolabah, and the other nearby in the same tree. The nest was about eleven metres above ground. Both birds were very alert and took to the wing while we were still a couple of hundred metres distant, and flew low overhead to investigate us before climbing to great heights to circle the area. The nest-tree was not approached, as every precaution was taken by us not to cause them to vacate the area. We'd been frustrated so often in the past, we weren't allowing ourselves any outward signs of optimism. Trips were taken daily along the nearby track, to unobtrusively confirm their presence in or near the nest-tree. We drove steadily past and didn't stop, thus they were never flushed. When, after a week, they were still there, we could no longer maintain a phlegmatic posture. We felt certain they were breeding or about to do so, but then we started to worry about egg collectors.

One of them had made it plain to us that the eggs were prized far above the $500 we'd been offering for information on a pair nesting. Only non-collectors, it would seem, would have been interested in the reward. We'd advertized throughout the nation, to no avail. Coincidently

Photographing the Grey Falcons' eggs.

The Strzelecki flood plain where the Grey Falcons nested.

The tower at the Grey Falcons' nest.

the offer had been made to rabbit shooters in the area, only three quarters of an hour before finding the nest two kilometres from their camp. We decided to keep the nest under surveillance till the eggs had hatched, though after a few days we were finding the daily vigil very boring.

On 29 August the tree was climbed and the four eggs photographed in the nest, and the opportunity was taken to do the little gardening necessary to give a clear photographic approach from the north. The Falcons attacked, cackling vigorously as they swooped close to the intruder. The female was quickly back on the nest as we moved off. Next day we set the tower up well back from the nest. It took exactly seven and a half minutes from the time we stopped the vehicle, till the moment we drove off, and there was a lapse of a further seven minutes before the female was back on the nest. We considered that quite acceptable, as the weather was mild to warm at that time. Had it not been so, we would not have attempted the task till it was. Two and a half hours later we moved the tower in to approximately ten metres from the nest, with an almost identical reaction - the nest being vacated for thirteen minutes. On the following day it was placed just under five metres away, and, getting no adverse reaction from the birds, the hide was entered. It was 1 September.

Within half a minute the female alighted on the edge of the nest and peered unblinkingly at the camera lens. It had been our intention to allow her to settle down before attempting any photography, but all those years of waiting for just this very moment proved too much. The camera shutter noise was enough to send her chattering off to do a few circuits of the nest tree before alighting again. She was very uneasy and cackled several times before taking to the wing again. In the hide the photographer sat motionless, the thrill of having the species on film was tempered by anxiety as to whether she would settle down. After what seemed an age, she returned and settled to brood her eggs. He glanced at his watch, it showed only seven minutes since the seeing-in party had left. He checked again. It

hadn't stopped.

She soon settled down and was joined by her mate a few hours later and some lovely photographs of the pair together were obtained. The pattern of behaviour that we'd observed from a distance, while keeping our vigil, was confirmed. The male would bring prey to the vicinity of the nest-tree, where the female would meet him. She would feed while he flew in to take up nest duty. On her return, he generally rose immediately and stepped aside to allow her to resume brooding. Occasionally, however, he seemed disinclined to give up his position and she would cackle till he did so. Sometimes he would alight on the nest with the prey and she would take it and fly to a nearby tree to eat.

Very windy conditions prevailed for several days and as a consequence only a few shots were taken. After a couple of such days, we found it necessary to remove further foliage that tended to spoil some shots by partly obscuring the subject. This procedure appeared to upset the Falcons far more than anything we'd done previously. When the hide was entered next morning the female seemed very nervous, even though she hadn't been any tardier than usual in returning to brood. Even the calls of Galahs and Corellas startled her, and she cackled quite often for no discernable reason. No shots were taken, nor the slightest movement made within the hide for the next half hour, in an

The clutch of four Grey Falcon eggs.

112

The pair of Grey Falcons at the nest. The slightly smaller male is on the left.

The female Falcon feeds her four tiny chicks.

endeavour to have her settle down. After the expiration of that time she took off, cackling as she went. She was not replaced by the male as was usual if she was feeding. As the minutes dragged by, anxiety mounted that we'd upset them to the point of desertion. Should the hide be vacated and the tower removed quickly? While contemplating the next move in a mood of deep despondency and self-recrimination for being such fools in attempting to work such a rare species on eggs, there came a short cackle and the female alighted on the nest and resumed incubating. She had apparently been feeding as there was blood on her bill. But why hadn't the male come in as usual to incubate in her absence? She had been away sixteen minutes. After a brooding session of ten minutes she again left the nest, cackling as usual. The male did not appear and she was away almost as long as the previous time. Something was wrong, as their behaviour was at sharp variance to that of the previous days. When she again left the nest a while later, the cameras were quickly packed and the hide vacated. It was, however, reassuring to be attacked as the tower was descended. Never before had attacking birds been so welcome. The rest of the day was spent observing them from our vehicle parked half a kilometre away. Their behaviour appeared normal, with both birds on or near the nest all day.

For the bulk of the incubation period the hide was entered daily with very little variation in the behaviour pattern of the birds. She would be back on the nest within moments of the seeing-in person's descent of the tower, sometimes before he'd reached the ground. Late in the incubation period she would sit tight till the hide was reached and then return as soon as the hide door was closed. The use of a seeing-in person was dispensed with and the behaviour pattern did not vary.

However, we became very anxious about the fertility of the eggs when they hadn't hatched by 15 September, when we took a short break and returned home. We thought the female was on eggs by 13 August, a week after we'd found them in occupancy of the nest.

When we returned four days later on 19 September, she was brooding four chicks which we estimated were only two days old. This indicated an incubation period of five weeks - a longish one for a bird of that size. As the chicks appeared to have hatched almost together it is possible incubation hadn't started till the clutch was laid, which could have meant it was something less than five weeks.

The behaviour pattern remained the same with chicks as with eggs - the male bringing prey to or near the nest. The female did all the feeding of the chicks, but the male brooded them while she was off feeding herself. She never fed on the nest even when he brought the prey to it. She tore off very small pieces and gave a cluck, similar to a domestic hen, each time she proffered a morsel, to attract the chick's attention. She was, however, very inefficient with the feeding of them. Each piece was proffered many times, but she seldom gave the chick time to accept it before proffering it to another. Often, after a while, she would eat the piece herself and tear off a fresh morsel. Even when a chick had taken a piece in its bill, she would retract it if it wasn't swallowed quickly. Her action was more one of nervous haste rather than impatience, as she spent considerable time feeding them. It was in complete contrast to most raptors which generally feed very young chicks slowly, gently and positively. Although the chicks only got about a fifth of that proferred during the first few days, it seemed to suffice.

After one feeding on the second day, the male, which had been present during the latter part of the feeding, began brooding the chicks. The female flew off, ostensibly to feed herself, but returned after a couple of minutes and cackled a demand for the male to allow her to resume brooding. Whether he just wanted to retain that chore for himself or whether he knew she hadn't eaten we could only speculate, but he refused to

The female Falcon burrows beneath the brooding male.

114

move. She flew off again and undoubtedly fed, as she had fresh blood on her bill when she returned a quarter of an hour later. He still refused to give up his position, and after a prolonged cackle she suddenly shoved her head below his breast and, pushing and squirming, burrowed her way under him, till he was brooding her and the chicks. He uttered no protest, not even the flutter of an eyelid, as he continued to share the nest with her for the next hour. Unfortunately the cine ran out of film as she moved to oust him, but several stills were taken of them in that unique position.

Before the eggs hatched, prey brought in to feed the female had appeared to be mostly small birds, like Richard's Pipits, but was now generally Crested Pigeon or Galah of which there were many in the area. Probably the preference for larger prey was due to having more mouths to feed. The brood was reduced to three a few days later when one died. Those three were thriving.

On 13 October, the hide was entered at 08:00. There was a stiff breeze blowing, but the tower was well guy-roped and the nest-tree was also roped to minimize nest movement. However, by 09:00 vibration in the hide made good photographic results improbable, so cameras were removed and the hide vacated. Wind velocity increased to gale force and we became anxious about the tower and the chicks' chances of survival. There was nothing that could be done while the storm raged, but by late in the day the wind had dropped somewhat and we were relieved to find the tower still standing. One chick was found dead on the ground a short distance from the nest-tree. As the tower had withstood the peak of the storm it was thought safe enough to climb. A quick ascent and a glance into the nest found the other two chicks crouched there. The nest was obviously that of a corvid, being deep and lined with feathers, but even so, we have no doubts that all the chicks would have been tipped out, had the tree not been guy-roped. Another corvid's nest near our camp, had been blown away during the day. We suspect that many such nests met the same fate that day, as they are usually in the topmost foliage of the Coolabahs, where they take a beating in such storms.

The nearby track was busy with road-trains hauling fifty to sixty tonne loads of pipes to the gas fields a few kilometres to the north. The track surface had become powdered, and huge dust clouds at times hid the nest from our view, but the Falcons seemed unperturbed by the noise and dust. Occasionally the male now helped with the feeding of the chicks, but for the most part he preened and dozed in the morning sun, while perched less than two metres from the camera lens. It was his favourite perch, and the proximity of the hide did not inhibit him in any way. He usually dozed while perched on one leg, the other hidden in the breast and belly feathers. The cine camera noise was enough to cause him to half open an eye when it started, but if kept going for a while, seemed to lull him to sleep. Both birds became very tame, and often stayed on the nest as the hide was entered or vacated, even though it was by that time, only a fraction over two metres from the nest.

After one very wet night, the track to the nest was negotiated with great difficulty, to rescue the cameras and flash heads left in the hide overnight. Two very bedraggled falcons looked pathetic as they huddled together brooding their chicks. They did not move as the hide was entered and the camera wiped dry enough to get a shot of them.

The female Falcon feeds a Galah to her chicks.

The male Falcon sleeps on his favourite perch near the nest.

The female Falcon shades her chicks with the male in close attendance.

The male Falcon with the two surviving, month-old chicks.

116

The fully fledged chicks at six weeks old.

Photography was almost impossible, as the lens fogged almost as quickly as one could dry it in the wet and humid conditions, so the hide was vacated for the day.

On 21 October we left the Strzelecki area to work on Peregrine and Black Falcons nesting near our homes.

Knowing the young Falcons would be close to flying, we left home on the evening of 1 November and drove all night to occupy the hide early next morning. It was a relief to find them still in the nest, but they would no doubt have flown if unduly disturbed. They had obviously been fed, and it was some hours before a very wary female alighted on the nest. She brought no prey, and the now hungry juveniles pecked at her talons. A very weary photographer, struggling to remain awake, decided to call it a day as he had obtained good shots of fully-fledged immature Grey Falcons.

The hide was entered before sunrise next morning, and after a two hour wait the female alighted with a small, partly plucked bird. She flew off as soon as one juvenile grabbed it. The chicks chased each other around the nest as the prey alternated between them. They had obviously not fed themselves before, and after a few unsuccessful attempts to do so, they lost interest in the prey, leaving it in the cup of the nest. We find this inability to feed themselves at this stage rather remarkable, although not unique, as the Black Falcons are somewhat similar. In contrast the

The juvenile Grey Falcons were remarkably similar to this recently fledged Gyrfalcon from Greenland. (Photograph courtesy of Prof. Clay. White, Brigham Young University, Utah).

Australian Hobby chicks we'd been working nearby, were feeding themselves when only half-fledged. The female returned to feed the young Greys later. Both adults had become more wary of the hide and we knew we wouldn't have time to woo them back to their former acceptance of us before the juveniles flew.

Female Grey Falcon.

Male Grey Falcon.

The two fully fledged chicks exercise their wings.

The female Falcon with her fledged chick.

We had planned to leave for home in the evening of 5 November, as day temperatures were reaching the mid forties celsius (over 110° F.) in the shade, and probably over 120° in the hide. The juveniles obliged by staying in the nest till 10:00 that day, when the one flew voluntarily. It alighted on the ground and ran toward a dead Coolabah, to flutter up and perch in it. A few moments later the female came into the nest and fed the other chick, which then flew off to join its nest-mate. It, too, alighted on the ground, apparently also lacking the confidence to make a landing directly into the tree.

From hatching to flight was exactly seven weeks, but we consider they were practically fledged at six. It was thirteen weeks since we'd found the Falcons in occupancy of the nest they subsequently used, and fifteen and a half since we'd found them nest-hunting. Coverage of their activities at the nest, from incubation onward, was more thorough than that of any other species to date. It would have been foolish of us not to have done so, after spending so much time and energy to find them in the first place. Regretfully we had failed to sound-record them, but with the knowledge gleaned, we feel that the second nest of the species should not be as hard to find.

A visit to the area in April 1980 found that the nest had been blown away. It was a little disappointing, as we were hoping that they might use the same nest again in the following season. We'd had reliable information that a pair had used the same nest in a Mallee tree in the Millewa, for several years during the 1960s, till the tree had blown down. Incidentally, it was quite close to the tree where we'd seen our first pair in 1974. That would indicate that the species is to some extent sedentary as far as their breeding area is concerned. Availability of a suitable nest would no doubt be the deciding factor as to breeding location. Our experience in this case could suggest that the general nesting area could be quite a large one, even when there are a lot of vacant nests about, as there was on the Strzelecki Creek in 1979.

In September, Richard Hollands - David's son -

and Alan Withers, a member of the Australian Raptor Association (A.R.A.), went into the Strzelecki area to search for the Grey Falcons on David's behalf. They knew exactly where we'd worked them the previous season. Surprisingly a pair of Grey Falcons flew into the tree while they were near it. Although the nest had blown down, the falcons apparently still had an attachment for that tree. It was indeed a remarkable coincidence that the birds visit coincided with that of the men, but there is perhaps the possibility that the men's presence near the old nest-tree attracted the Falcon's curiosity, causing them to fly in from some distance away.

Late in October 1980 another two A.R.A. members went into the Strzelecki area. David Baker-Gabb and Dr. Jack Pettigrew were looking chiefly for Letter-winged Kites, as Jack, an eye specialist researching avian vision, was keen to study that of the night-feeding kites. They saw four Grey Falcons, possibly last years brood and parents. From their description, two were juveniles, darker on the back and breast than the others, but having the yellow cere and legs, which they don't have when they first leave the nest. If they were last season's young, then the cere and legs become yellow during the first year. In the interests of our studies we now realize we should have colour tagged them in the nest.

The chick about to take off on its first flight.

120

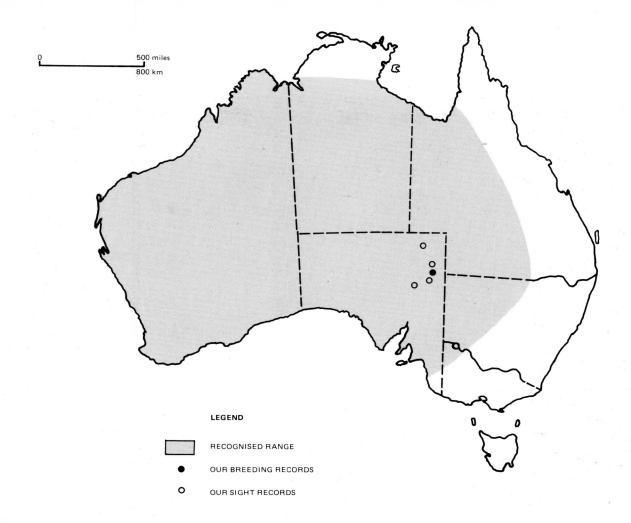

LEGEND

▢	RECOGNISED RANGE
●	OUR BREEDING RECORDS
○	OUR SIGHT RECORDS

GREY FALCON *Falco hypoleucus*

falco- falcon (L); *hypo-* under (Gk); *leucos-* white

OTHER NAMES: Blue hawk; Smoke hawk.

LENGTH: 340 - 430mm; Female larger than the male.

WINGSPAN: Approximately 950mm.

DISTRIBUTION: Uncommon to very rare; frequents open, drier areas of mainland Australia, particularly along timbered watercourses of the interior. Probably largely nomadic but sometimes sedentary. Not found outside Australia.

VOICE: Hoarse chatter **'chak-chak-chak-chak'**, generally not as rapid as the Peregrine Falcon. Also **'cluck-cluck'**, probably as contact call.

PREY: Chiefly birds, but also small mammals, reptiles and insects. We have recorded only birds, usually unidentified, but those recognised were Crested Pigeons and Galahs, while smaller birds, in some instances, were Pipits.

NEST: The deserted nest of another raptor or corvid is used, sometimes being used for a number of years. The nest we found was that of a corvid, in the top of a tree, 11 metres above ground.

EGGS: Two to four 51 x 38mm; they are rounded ovals of pinkish buff, heavily spotted with red-brown. The one clutch we have found contained four eggs. Eggs are laid from July to November; our record being Mid-August. Incubation appears to be 32 to 35 days and the chicks fledged in about 42 days.

AUSTRALIAN KESTREL
Falco cenchroides

A Kestrel alights with a skink in its bill.

This beautiful little falcon, Australia's smallest, is very common throughout most of mainland Australia, though rarer in the north and in Tasmania. It frequents a wide variety of habitats, being most common in open woodlands and cultivated areas, but largely absent from densely forested country. It is often found in urban and suburban parks and gardens. The Australian subspecies is occasionally found in Norfolk and Lord Howe Islands and in New Zealand. Another subspecies is found in New Guinea and neighbouring islands. The Australian Kestrel is also known as the Windhover and Sparrowhawk but is probably best known as the Nankeen Kestrel, a name derived from its chestnut coloration. Although the size difference is sometimes detectable in the field, plumage is more diagnostic of sex. Females are more heavily spotted and streaked on the back and have chestnut tails barred black, while the males tail is an off-white or pale grey, tipped white, with a broad black band near the tip.

The Kestrel in flight.

122

A pair of Kestrels feeding their young. The chestnut tail of the female contrasts with the pale grey tail of the male.

The scientific name for the bird, *falco* meaning 'falcon', *cenchris* for 'speckled hawk', and *eidos* meaning 'form' or 'like', suggests a 'falcon which is like a speckled hawk'.

Probably nowhere in Australia is the Kestrel more common than around our home territory in north-western Victoria. Many a working day has been interrupted to watch and marvel at a hunting bird, for this species is a perfect hoverer, hanging on the wind with gently flapping wings and fanned tail, before diving steeply to the ground to take a mouse or insect, which is then taken to a nearby perch to be eaten. Kestrels also take prey spotted while perched on a high vantage point. Proof of their extraordinary powers of vision is ably demonstrated when one will leave its perch, fly directly to pick up a mouse, one hundred or more metres distant, and return to its perch to feed. Apart from mice, which appear to be their favourite food when available, they take other small mammals, such as kitten rabbits, small reptiles and birds, spiders and insects.

Kestrels don't build a nest of their own, most often using an open hollow in a tree or a hollow limb, or alternatively they appropriate the old stick nest of another species, usually that of another small raptor or corvid. They will also nest on rock ledges on both coastal and inland cliffs and rocky outcrops; and occasionally even on ledges on city buildings. In 1974-75, along the Birdsville Track we found that they mostly nested in stick nests, due, no doubt to the shortage of suitable tree hollows. In our home area, however, hollows were generally chosen, so it was with much puzzlement that we kept checking eminently suitable hollows in a tree from which a pair of Kestrels were seen to fly, only to find no sign of a nest. The only stick nest in the tree was that of Chestnut-crowned Babbler, which, with its enclosed top and small side entrance was not considered suitable, until on passing the tree for the fourth time, a Kestrel was seen to leave from the nest or very close to it. An inspection of the nest revealed that the Kestrels had widened the side entrance, leaving the roof intact, and now had a clutch of four eggs. A few days later a similar nest was found nearby and a third was found on our property. We erected a tower at this latter nest and found that the side entrance was used till the eggs hatched, the nest then being opened out. When the chicks in that brood had flown, the Kestrels immediately relaid, rearing a further three chicks. At this nest the incubation period was 26 - 28 days, with the chicks fledging in 26 days.

In 1978 we set up at an open stick nest containing three chicks, about two weeks old. In the nest, we were somewhat surprised to find a kitten rabbit about the size of a native rat. Part of this was fed to the chick then the remainder removed by the female. Later she called to the male who brought grasshoppers which she fed to the chicks.

Four days later at 08:30 the hide was again entered. There was no action till seventeen minutes later when the female brought a grasshopper which was quickly snatched and eaten by one of the chicks. That marked the beginning of the most hectic activity we have ever witnessed at a nest. Over the next forty minutes the pair of Kestrels brought in prey fifteen times. There was a lull in activity for half an hour, with the female perched on a branch above the nest and the male in a tree nearby. Then suddenly they were hunting again, twenty-seven visits being made in the next hour and a further thirty-one in the ensuing one and a half hours, to make a total of seventy-four visits in three and three-quarter hours. The prey was mainly spiders and insects such as crickets, dragon flies and grasshoppers, but also four skinks - three of which were brought by the male. He appeared to visit the nest about half as frequently as the female, but was possibly doing most of the hunting, passing the prey to her to bring to the nest. When one considers the number of insects Kestrels must account for in a season, this statement about the Kestrel, from

The male Kestrel visits the nest, the abandoned one of a Brown Goshawk.

A female Kestrel with a piece of rabbit for her chicks.

Brown and Amadon's **Eagles, Hawks and Falcons of the World:** 'Generally a harmless and beneficial bird which should not be persecuted', must surely be the understatement of the century.

A second pair were nesting just a few kilometres away, this time in an open hollow in a mallee tree, just three metres from the ground. This nest contained four eggs when found and the Kestrels behaved as we had come to expect, circling overhead, chattering their rapid **ki-ki-ki-ki,** but showing no other signs of aggression. However, their behaviour was totally different and unexpected when we returned two weeks later to erect a hide.

Both birds were much more vocal, and as we approached the nest-tree they stooped constantly to within a few centimetres of our heads.

On checking the nest it was found to contain three chicks about one week old, so three large fruit crates were stacked one on the other about four metres from the nest. This work was done with a constant ducking of heads as the Kestrels kept up their attack. The crates reached to nest height, leaving only the hide to be placed on top to complete the job. However by then the birds were even more excited. As the hide was carried we endeavoured to keep it between our faces and the

The female Kestrel, with a cricket for her chicks, perches above her nest.

attacking birds because it seemed inevitable that sooner or later we would have our faces raked by their razor sharp talons. The Kestrels countered this strategy by streaking across the sky till they were almost directly overhead, then locking up their wings, they would come plummeting down almost perpendicularly, missing the hide by mere millimetres, and our foreheads by little more. It was with no small measure of relief that we finished placing and securing the hide and left the area.

On subsequent visits to work these birds we found them fairly aggressive, right up till the last chick flew from the nest, but they always settled down after the hide was entered and kept a constant supply of prey for the chicks, once again largely spiders and insects, but also a few kitten rabbits. Unfortunately all three chicks developed eye infections, one bird having both eyes completely closed in much the same manner as a rabbit with myxomatosis. This bird was eventually found dead beneath the nest, while the other two, each with one blind eye, fledged and left the nest.

A typical clutch of Kestrel eggs in a nesting hollow.

The female Kestrel at her nest hollow in a mallee tree.

Our improvised "tower" at the Kestrels' nest.

The female Kestrel brings a mouse for her brood.

A fully fledged Kestrel chick in its nest hollow.

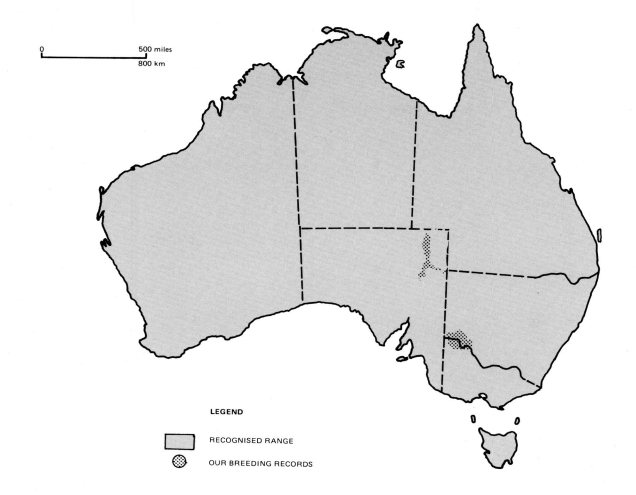

LEGEND

▭ RECOGNISED RANGE

⊗ OUR BREEDING RECORDS

AUSTRALIAN KESTREL *Falco cenchroides*

falco - falcon (L); *cenchris* - speckled hawk (Gk); *eidos* - form, like (Gk).

OTHER NAMES: Nankeen Kestrel; Windhover; Sparrowhawk.

LENGTH: 300 - 350mm. Female larger than the male.

WINGSPAN: 700 - 750mm.

DISTRIBUTION: Common in many habitats throughout most of Australia, less common in Tasmania and northern Australia. Favours open woodlands, grasslands, agricultural areas, urban and suburban parks and gardens; absent from dense forest. Either sedentary or nomadic. The Australian subspecies *F. c.cenchroides* is found in New Zealand, Norfolk Island, Lord Howe Island. Also found in New Guinea and neighbouring islands.

VOICE: A high-pitched, shrill chattering '**ki-ki-ki-ki ...**', or a rapid twitter.

PREY: Insects, spiders, small mammals, small reptiles, small birds. We recorded kitten rabbits, mice, skinks, spiders, and such insects as crickets, locusts, dragon flies and beetles.

NEST: Most often a hollow in a tree is used, but also the stick nest of other species; rock ledges, or even ledges on city buildings.

EGGS: Usually three to five, rarely six, 38 x 31mm. They are rounded ovals, smooth, slightly glossy, pale buff usually well-covered with spots and blotches of red-brown and dark-brown. Of the 21 clutches we found: 1 contained six eggs, 7 had five, 10 had four, and 3 had three eggs. They are usually laid from late August to late November. The incubation period is 26 - 28 days and chicks fledge in about 26 days.

BROWN GOSHAWK
Accipiter fasciatus

This species is known by several other names, although perhaps the most common alternative is Australian Goshawk. It is well-represented in most areas of the mainland and Tasmania, and extends outside Australia, mainly on islands to the north and as far away as Fiji. There are two subspecies: *A. f. didimus* mostly found in the northern coastal regions, and *A. f. fasciatus*, generally in the south. The latter may migrate north in winter.

The behaviour of this species towards human intruders near the nest varies widely. Some birds fly furtively away from the nest when it is approached, while others attack vigorously. For years this was the only raptor species that had struck us to the point of drawing blood.

In October 1974 we checked a nest which we knew to have been used by Goshawks the previous year, and were delighted to find the birds incubating four eggs. We erected our tower about eight metres from the nest and within minutes the female Goshawk returned. After taking a few photographs we decided to leave the tower standing so we could do some further work next day. However, when we returned we found the

nest empty, though the birds were still in the area. We later learned that a local egg collector had taken the eggs.

A few weeks later we found the birds had laid again in a nest only one hundred metres from the first. We immediately returned home to pick up our tower, but when we returned two hours later, we found the nest empty. Once again we were thwarted by the egg collectors.

Although those first Goshawks showed no aggression toward us, they could, in view of the treatment they suffered, be excused less sociable behaviour in subsequent years. We were unable to find them nesting again, but some of their kin certainly did their utmost to avenge them.

On 11 November 1975 we found a nest with three chicks which we estimated to be about ten days old. The stick nest was on a horizontal branch of a Belar tree, at a height of ten metres. We set up about eight metres from it. Climbing the tower was quite an ordeal, as both birds would attack fiercely and would strike if one didn't turn and face them. As it was impossible to enter the hide without turning one's back on them, it was inevitable that legs would be scratched. After that

A female Brown Goshawk with chicks.

Brown Goshawk in flight.

An attacking Brown Goshawk.

An aggressive brood of Goshawk chicks.

experience, long trousers were the order of the day. The female tore threads from the seat of one pair, luckily without damage to the wearer's buttocks.

On one occasion the back of a hand was raked as it was extended to focus the camera lens, and on another the female ran around on top of the hide for about a minute, screeching loudly. One wonders what would have happened had she found her way into the hide with the photographer, for she was without doubt the most irate female we'd ever seen. However, with the departure of the seeing-in party, both birds would quickly settle down and go about their parental duties of hunting, feeding and brooding.

The day's routine started with the female collecting a spray of eucalypt leaves to spread on the nest. She would then call once or twice from the nest and soon afterwards fly out to a nearby tree to collect prey from the male. At this nest it was nearly always a half-grown rabbit or a part of one. On rare occasions a bird varied the diet.

He seldom came to the nest and when he did only stayed momentarily. She fed the chicks till they ignored the proffered pieces and they usually dozed for awhile, but it wasn't long before they would be looking to be fed again. They got through an amazing amount of food, probably more than any other raptor of comparable size.

The female was markedly larger than the male, probably twice his weight. He would be very hard to distinguish from the female Collared Sparrowhawk which he resembles in size and plumage.

In November 1979 a nest with two eggs was found in an area where a pair had been nesting for some years, indicating a sedentary disposition. It was close to home and could be monitored at will. There was a third egg when inspected two days later.

Wishing to study the roles of the adults during egg incubation, we set up tower and hide almost immediately at roughly ten metres from the nest. After noting that it was accepted within half an hour by the female returning to the nest, we made

The male Goshawk about to leave the nest as the female approaches.

no further visit for a fortnight. The tower was then moved in to 3.8 metres from the nest and the hide entered. The female had left quietly as the tower was approached and she was back on the nest a quarter of an hour after the seeing-in party left. She made a noisy approach and perched above the nest voicing alarm for a minute before dropping into the nest to resume brooding. She sat quietly for five hours without much movement except occasionally half rising to move the eggs slightly with her bill before settling again. At 12:45 the male called a few times from nearby and she left the nest. A few moments later he alighted and immediately settled over the eggs. He was slightly darker in plumage generally and the iris of his eyes a deeper yellow than the female's. His smaller stature was very pronounced, especially when brooding. An hour later she made a few calls from nearby and he left the nest to be replaced by her. This pattern of behaviour varied only by an hour or two as to when the change-over was first made each day. Sometimes it came as early as 11:00 or as late as 13:00. There appeared to be only the one change-over for her to feed and her feeding time varied between a half and two hours duration while on eggs. Sometimes either bird would alight on the nest before the other had departed and that was one of the main

The female Goshawk stands to leave her nest of three eggs.

130

reasons we sat for such long periods each day, to get a shot of them together on the nest. Eight-hour sessions were not uncommon. Many thought we were bird-brained and we were sometimes inclined to agree, especially after very long sessions when each bird had called the other off before coming in to the nest. On most occasions that we did have both together they only stayed momentarily and they were not always positioned satisfactorily, being one behind the other, or one being on the front of the nest with the other on the back of it, making it impossible to focus on both at once.

A typical clutch of Brown Goshawk eggs.

On 20 December there were two freshly-hatched chicks, quite lively and chirpy. No attempt was made to feed them till late in the day when the female left the nest and was replaced by the male. She soon returned with a piece of well picked-over hind quarters of rabbit. She tore off some small morsels but after appearing to study each piece while held in her bill, she must have decided they weren't baby food and ate them herself. After a few more pecks at the prey she decided none of it was suitable and flew off to dump it. The photographer mentally agreed with her, it was hardly suitable for crow bait. After a short brooding session she left again and returned with the same pieces as before. After trying unsuccessfully to get something suitable for the chicks she started where the contents of the stomach had spread onto the fur of the upper hind legs. Once again she was about to proffer this mess to the chicks but hesitated and ate it herself; she must have been very hungry! Eventually she managed to get a few morsels that were reasonable fit for tiny, newly hatched chicks. It could not by any stretch of imagination be called a meal.

A visit two days later found only one chick and the third egg still in the nest. One chick had apparently died and its body been removed from the nest. At 10:30 the female raised herself from a brooding position and peered for a few moments

The female Goshawk removes the egg shell as soon as the chick has hatched.

The smaller male arrives to relieve his mate.

A juvenile Brown Goshawk.

into the nest, before carefully backing to the nest edge to watch the third egg hatch.

From our observations it would appear that incubation started after the second egg was laid. There were two eggs in the nest when found on 19 November. Two had hatched either on the night of 19 December or the next morning, the third two days later, thus incubation appears at thirty-one or thirty-two days.

While searching in 1980 for a nest of Collared Sparrowhawks in the Millewa in north-west Victoria, and on Tapio Station in south-western New South Wales, we found many goshawks nesting. Most nests were in belar trees growing in clumps ranging from a half to perhaps two hectares in area. On Tapio the belar stands occupied the higher ground along sand rises and were separated by open areas of dry grassland. Rabbits were in near plague numbers and no doubt formed the greater part of the goshawks' prey. The close nesting of the species was also probably influenced by the abundance of prey.

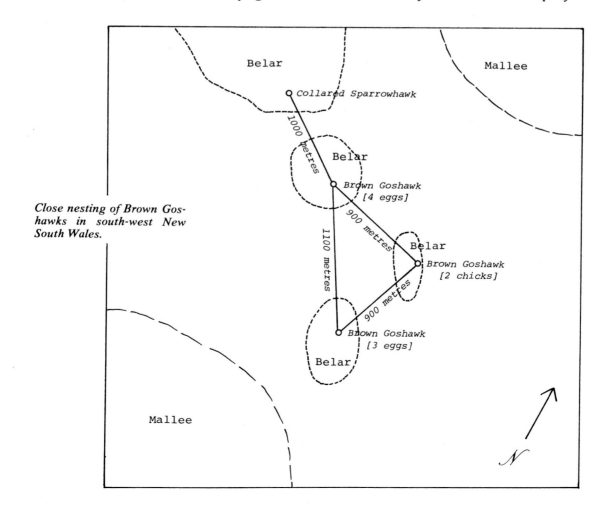

Close nesting of Brown Goshawks in south-west New South Wales.

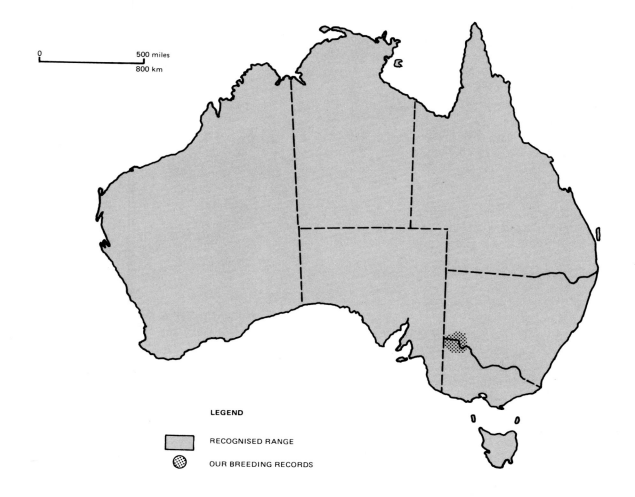

LEGEND

RECOGNISED RANGE

OUR BREEDING RECORDS

BROWN GOSHAWK *Accipiter fasciatus*

accipiter - hawk (L); *fasciatus* - banded (L).

OTHER NAMES: Australian Goshawk; Chicken hawk; Grey-headed goshawk; Western goshawk; Collared goshawk.

LENGTH: Female 430 - 560mm; Male 370 - 430mm.

WINGSPAN: 750 - 1000mm.

DISTRIBUTION: Common in a variety of habitats throughout Australia; favours woodlands and open forest, particularly near water. Mostly sedentary. Two Australian subspecies. *A. f. didimus* - probably restricted to coastal northern Australia. *A. f. fasciatus* - generally southern Australia; may migrate northwards in winter. Also from Christmas Island, through Timor and New Guinea to Fiji and neighbouring Pacific Islands.

VOICE: Alarm call is a rapid **'ki-ki-ki-ki'** or **'kek-kek-kek-kek'**, the female having a deeper call. Also a slow **'seep-seep-seep'**.

PREY: Rabbits, birds, reptiles and insects. We recorded mainly rabbits with an occasional small bird (passerine).

NEST: A rough structure of sticks and twigs, usually with leaves throughout, lined with green leaves, generally placed in a horizontal fork from six metres to over thirty metres above ground. Nests we measured usually ranged from 500 - 700mm in diameter and from 250 - 300mm deep. We have, however, worked at a nest that was only 380mm in diameter and 180mm deep, while other nests in tall forest trees appeared considerably larger than those measured. The latter were used for a number of years in succession, and probably added to each season.

EGGS: Clutch size can range from one to five eggs, but is usually three or four, measuring 46 x 37mm. They are bluish-white, smooth but glossless, can be unmarked or sparsely marked with spots and streaks of red-brown or lavender. We recorded 10 clutches: 5 with three eggs, 4 of four, and 1 of two. Eggs are laid from September to December, our records indicating egg-laying in October and November. The incubation period is 30 - 33 days and the chicks fledge in about 30 days.

COLLARED SPARROWHAWK
Accipiter cirrhocephalus

This species is found in most timbered areas of Australia and subspecies extend to New Guinea and neighbouring islands. There are two subspecies in Australia: *A. c. cirrhocephalus* over most of the mainland and Tasmania, and *A. c. quaesitandus* which inhabits the extreme north of the continent. It is not a common bird, and certainly far less common than the Brown Goshawk, which it resembles.

There are many similarities between this species and the Brown Goshawk. The female Sparrowhawk is almost identical in size and plumage to the male Goshawk, the discernable difference being the longer middle toe of the Sparrowhawk. In flight the shorter, square-cut tail of the Sparrowhawk usually distinguishes it from the Goshawk, which has a rounded tail, although if the Goshawk has lost some tail primaries its tail will also appear rather square. In the area near our homes the two species prefer the same type of habitat, areas of Belar, increasing our difficulty in identifying them. Sometimes we were uncertain which bird we had found until we began working them. Both species have a great range of reactions to human intruders near the nest. Some attack vigorously, others placidly accept the intruder, although the Sparrowhawk is generally easier to work.

In November 1975 we found a pair of Sparrowhawks nesting in a belar tree on Tapio Station in south-west New South Wales. The nest contained three eggs, and as the brooding female was very reluctant to leave, we decided it would be safe to attempt some photography.

A typical clutch of Sparrowhawk eggs.

The tower was erected only two and a half metres from the nest without the brooding bird being disturbed. Indeed, when we began trying to photograph her, we had a frustrating and boring time, as she sat low in the nest, hour after hour.

Finally, hoping to photograph her standing on the edge of the nest, a long stick was used to gently prod her into a standing position. However this tactic failed, as the determined bird simply hopped over the stick to settle down onto the eggs once more.

In contrast, after the eggs had hatched, both adult Sparrowhawks would attack vigorously whenever we approached the nest tree, with the female grazing us as we climbed the tower.

During the brooding period the male did all the hunting. Prey was almost exclusively small birds, though there have been reports of the species taking much larger ones such as Grey Teal. Per-

In north-west Victoria and south-west New South Wales nearly all Sparrowhawks' nest we found were in patches of belar.

A female Sparrowhawk watches over her small chick.

haps, but the possibility of the hunting bird being wrongly classified should not be ruled out.

During a very hot spell it was fascinating to observe the female's efforts to ameliorate conditions for her three chicks, even though the nest was in the shade. She would stand over them for hours with her tail spread and wings extended to direct what little breeze there was over them. The involuntary upsweep of her tail when the breeze freshened occasionally, was her only movement.

She shades them with her body and outstretched wings

The species was seldom sighted between 1975 and 1980, although being busy on other species we made no great effort to hunt for them. The areas they'd occupied a few years ago we now found to be taken over by Goshawks. In one case, even the nest tree had been taken over, though the nest had been replaced.

Late in November 1980, after considerable searching, during which we found many Brown Goshawks, we found a nest with two chicks possibly a fortnight old as wing and tail primaries were appearing. The female proved to be the placid type, just watching from a nearby tree as the tower was erected a little over two metres from her nest. Like a lot of nests of the species seen or worked by us in the past, it was on a horizontal branch of a belar. She was back on the nest shading the chicks within minutes of the seeing-in party leaving. Occasionally she gave a few calls, presumably for the male to hustle up a meal as it wasn't long before he was back in a nearby tree with a small bird, probably a Thornbill of some kind. She flew out to snatch it from his

talons, chatter briefly while doing so and headed back to the nest. Within three minutes it had been fed fairly evenly between the two chicks. After cleaning the last vestiges from her talons she gave a few more calls, probably to let the male know they could do with the same again. For the next two hours she stood shading the chicks, occasionally taking a short nap while still maintaining her stance above the chicks. Getting no response from her intermittent calls to the male, she apparently spotted him perched without prey in a nearby tree and flew out to chatter sharply at him before returning to the nest. Her actions had the desired effect as it wasn't long before he was back at his perch with another small bird that could have been a female Red-capped Robin.

By our third visit, the Sparrowhawks took very little notice of us. The chicks were well-feathered, but still had considerable down to shed. The disposal of the small birds now took less than two minutes as the female tore off larger pieces to feed them.

By 3 December 1980 the young birds were

At another nest, the female Sparrowhawk feeds her two chicks.

approximately four weeks old and it would have caused no surprise if they had flown as the hide was entered. Their wing exercizes had them airborne on numerous occasions, but they always managed to alight back on the nest. One would walk out on a limb toward the hide until it was only 1.5 metres away.

The male brought prey to a nearby tree almost hourly where the female took it and brought it in to feed the juveniles. They sometimes grabbed it from her and pecked at her when she sought to retrieve it, but she always won out and fed it to them. As the weather was mild she didn't stay to shade them, but flew to a nearby tree immediately the prey was finished. The nest was just a flimsy platform now, having been trampled flat. The relining with fresh green leaves had probably stopped about the time the young started to fledge as the leaves in the nest when first found were not fresh and none had been added since we'd had it under observation. The young put plenty of pep into their wing exercizes and had to clutch tree branches near the nest to stay 'grounded'. The bulk of their time was spent

The square-cut tail of the Sparrowhawk is evident as the female shades her chicks.

preening and the down-covered twigs and branches on the lee side of the nest bore testimony to their efforts in that regard.

The two fully fledged chicks on the small platform nest. Some of their recently shed down clings to the tree.

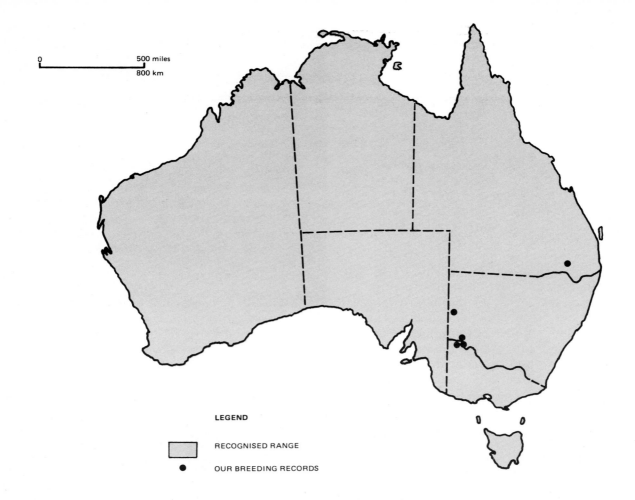

LEGEND

▭ RECOGNISED RANGE

● OUR BREEDING RECORDS

COLLARED SPARROWHAWK
Accipiter cirrhocephalus

accipiter - hawk (L); *cirros* - tawny (Gk); *cephale* - head (Gk).

OTHER NAMES: Australian Sparrowhawk; Chicken hawk.

LENGTH: Female 360 - 390mm; Male: 280 - 330mm.

WINGSPAN: 600 - 760mm.

DISTRIBUTION: Uncommon to moderately common in most timbered areas throughout Australia; sedentary. Two Australian subspecies: *A. c. cirrhocephalus* most of Australia except the far north. *A. c. quaesitandus* northern Australia. Other subspecies extend to New Guinea and neighbouring islands.

VOICE: Alarm call is an extremely rapid and shrill **'ki-ki-ki-ki-ki'**. At other times a fairly slow, mellow **'swee-swee-swee'**.

PREY: Almost exclusively birds, usually small, but larger birds, such as grey teal, parrots and cuckoo-shrikes have been reported; as have small mammals, lizards and insects. We saw only small passerines brought to the nest.

NEST: A shallow platform of light twigs, lined with green leaves, placed on a horizontal branch 6 - 30 metres above ground. A nest typical of those we've worked, measured 300mm in diameter and was 125mm deep and 8 metres above ground.

EGGS: Usually two to four, 39 x 31mm. They are rounded ovals, bluish-white, without gloss, unmarked or sparsely spotted and blotched with red-brown. We have recorded five clutches: 2 contained four eggs, 2 contained three, and the other only two eggs. Eggs are laid from July to November; our records indicate egg-laying from late September to late October. We have been unable to record incubation and fledging periods accurately, but they would appear to be approximately three weeks and four weeks respectively.

GREY GOSHAWK
Accipiter novaehollandiae

This Goshawk occurs in two distinct colour phases - grey and white. The grey phase is grey above and white below, while the white phase is pure white. Not surprisingly, the alternative name, White Goshawk, is used as often as the official name.

Though reasonably common where suitable habitat exists, the species is generally not often seen, because it frequents rainforests, woodlands and well-timbered watercourses. The two phases tend to dominate particular areas. The white phase is moderately common in Tasmania, and dominant in the south-east and north-west of the mainland. The grey phase is dominant along the east coast. A number of subspecies occur through New Guinea and neighbouring islands, though most bear little resemblance to the Australian forms, being generally more like our Brown Goshawk.

The shyness of the Goshawk has meant its nests are often in heavily-timbered country, and difficult to find. In 1975 we studied a white phase bird interbreeding with a Brown Goshawk*, and prior to 1978 had only one other sighting, a brief glimpse of another white phase bird near Maryborough, Queensland.

A captive grey phase Goshawk, photographed in David Fleay's Fauna Reserve, Queensland.

★ *See following chapter: 'Interbreeding of Grey and Brown Goshawks'.*

In August 1978, while taking a well-earned rest on the banks of the East Alligator River in Northern Territory, we were alerted by the alarm calls of small birds in the leafy canopy above us. We looked up to see a beautiful pure white female Goshawk alight ten metres above our heads. She remained watching us for five minutes until we reluctantly moved on in search of the Red Goshawk. Several days later, and some kilometres away we got our first sighting of the grey phase of the species. Again it was a female bird and she too appeared to be doing some man-watching, peering down at us from her vantage point in a tall Melaleuca. She was perhaps not as striking as the White Goshawk, but was still an extremely handsome bird.

Mike Traynor, Assistant Curator of Birds at the National Museum, Melbourne, gave us directions to a nest near Apollo Bay, which he said was in a deep gully, but quite close to a road. Though Mike felt we could use a long lens to photograph the nest from the road, we doubted that approach would give us the type of photographs we were seeking.

David Hollands, who had inspected the area some time earlier, agreed that the view from the road was unsatisfactory. He also felt that it would be impossible to work at all, because of impenetrable blackberries growing on the steep sides of the gully. We felt, however, if the nest was so close to the road, then we must be able to clear a path to it.

We left from Melbourne for Apollo Bay early on 9 October travelling via the Great Ocean Road. The township is situated on the coast one hundred and eighty-five kilometres west of Melbourne and the picturesque Otway Ranges form a lovely backdrop. About a kilometre from the edge of the town we searched what we believed was the nest area, without success. After trying unsuccessfully to contact Mike again by phone we put our problem to the local postmaster in the hope that he might know someone, a local bird enthusiast perhaps, who knew of the nest. What luck! Not only did Bill appear to know everybody and most of their activities, he also knew the nest personally and took us through the back door of the post office to point out the exact location of the nest-tree in a deep gorge-like gully in the hills behind the town. We had been searching the right area, but the added information enabled us to locate the nest in a few minutes.

The rough terrain was nowhere as formidable as we'd been expecting, and we had the added advantage of a recently constructed fence on which to cling as we descended to the small stream

The White Goshawk's nesting area, in a dense clump of trees in a gully near Apollo Bay.

about fifty metres below road level. The nest was on a horizontal branch of a huge eucalypt and about twenty-five metres above the stream bed. It was noticed that a branch on the opposite side of the tree to that of the nest reached right to the bank. By clambering thirty metres up that bank it was possible to get onto the branch and walk, tight-rope fashion, to the centre of the tree and then along an opposing branch to the nest which contained two eggs. We were elated, but quite out of breath by the time we'd climbed back to our vehicle. We would only need one of our regular hides placed in the nest-tree from which to film. It would, under the circumstances, prove much easier than using a tower. We returned home.

Leaving again on 30 October at our usual time, we arrived at Apollo Bay at noon. The weather was cold and windy, but we still raised a sweat getting our gear down to the stream bed below the nest. We had to work fast once we started to prepare the platform for the hide near the centre of the tree, slightly above nest level and eight metres from it. The job was done in three half-hour shifts with one hour intervals to allow the female back on the nest and so obviate any risk of the eggs getting too cold. After completion of the platform the hide was hauled up and placed in position. The female behaved admirably and at one stage returned to the nest while the work proceeded eight metres from her.

That night we camped on the beach with the waves rolling to within a few metres of us. We slept soundly, a direct result no doubt of energy expended climbing up and down those steep banks at the nest site, and the nervous tension that is always present to some degree while working

Hauling up the hide.

The Apollo Bay nest with its small clutch of two eggs.

The female Goshawk broods her eggs.

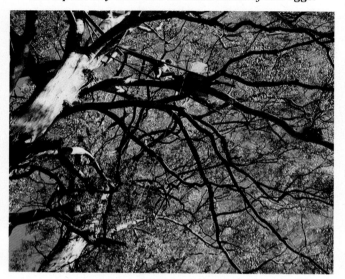

Positioning the hide twenty five metres up.

Entering the hide for the first close-up shots.

The hide. less than four metres from the nest.

in the nest-tree. We were fortunate that it did not rain during the night as we'd barely finished our breakfast when it started. It did not augur well for a start to our filming, and it didn't make those steep banks easier to negotiate, on the contrary. However, we did obtain some reasonable film of the hide preparations as well as the birds' activities at the nest. We felt so happy with the results of our efforts that we treated ourselves to the relative luxury of an overnight van. It was well we did as it rained heavily for much of the night.

We started early next morning and got some good results as the weather improved. One egg had hatched overnight and we were able to film the chick getting its first feed taken from a small well-plucked bird. The male brought fresh leaves to the nest and did a short spell of brooding in the female's absence. He was, however, very wary and wasted no time vacating the nest on her return. She showed no fear of the hide and appeared to be acting quite naturally, ignoring the photographer entering and leaving it most of the time. We left the hide **in situ** for our next planned visit in about ten days.

The female Goshawk gives her chick its first feed - from a well-plucked bird.

The smaller male at the nest while the female is away feeding.

On 9 November we left at 20:00 and 'camped' south of Ballarat. After four hours sleep on the roadside, in 'full marching order', we arrived at the nest site at 08:00. The weather was no better than on the previous trip; rain on the ranges, cold and bleak along the coast. No time was wasted getting into the hide to do a little more cine before shifting the hide closer to the nest. We had come prepared with some short lengths of aluminium angle to be secured to the branch, and on which we planned to place the hide. Manoeuvring the hide along that branch and onto those angles was a ticklish job, but was accomplished without mishap. Speed was essential for the chick's sake, caution for our own. No photography was attempted that day from the new position only 3.8 metres from the nest.

The night was spent on the foreshore. The howl of the wind in the nearby pines and the roar of the surf failed once again to disturb us. By 08:00 we were making our way down the bank to the nest-tree. It was warmer down there, but the sight of a huge branch that had broken from the nest-tree overnight reminded us of the potential dangers inherent in that type of eucalypt. It is unwise to camp beneath them at any time, but more so in the summer, when after a hot day large branches may crash without warning as the temperature drops in the evening.

We managed to get some good shots from that close position as well as some equally good cine of their behaviour. Our hide was left *in situ* for a planned final visit in a fortnight. There was little fear of it being interfered with in our absence, situated as it was in a position accessible only to the most competent climbers, or the most foolhardy.

The female feeds the eleven day old chick. The unhatched egg is still in the nest.

143

By 25 November the chick was almost fledged, but still had a fair amount of down to shed. We estimated it would be at least another week in the nest which put fledging at between five and six weeks. The rotten egg was still in the nest that day, but was gone by the following morning. Only the female visited the nest, and those visits were most infrequent being at roughly four hour intervals. The chick, however, did not appear hungry during those intervals and occupied itself preening and wing-flapping between naps. That period was one of intense concentration for the photographer trying to get action shots of the female alighting on the nest. Those shots require very precise timing and for one to remain alert enough to perform that split-second action is often quite exhausting both physically and mentally, especially when the bird's visits are so infrequent as this one's were. Most of that weekend was spent in the hide for very little photographically as the female made only four visits in that time. The weather was very hot and sultry and conditions in the hide most oppressive.

We lowered our hide down by rope whilst working from the creek bed twenty-five metres below. After releasing the hide, the rope became snagged on the nest branch, so we left it there, as $10 worth of rope hardly seemed worth the effort and risk to climb that bank and negotiate those branches again to retrieve it. No doubt we'll have another go at working the nest some time as no matter how good the results are, one feels they can always be improved.

To find that nest and cover it as we did had meant four trips, each of thirteen hundred kilometres and involved at least eight long days of effort, but the results certainly justified it all.

The twenty six day old chick.

The female glances at the hide before feeding the chick.

A view of Apollo Bay from the nest site.

RECOGNISED RANGE

● OUR BREEDING RECORDS

○ OUR SIGHT RECORDS

GREY GOSHAWK *Accipiter novaehollandiae*

accipiter - hawk (L); *novaehollandiae* - of New Holland

OTHER NAMES: White goshawk; Grey-backed goshawk; New Holland goshawk; Rufous-breasted godhawk; Variable goshawk; Vinous-chested goshawk.

LENGTH: 405 - 530mm. Female much larger than the male.

WINGSPAN: 710 - 1005mm.

DISTRIBUTION: The Australian subspecies *A.n.novaehollandiae* is uncommon to moderately common in rainforest, woodlands and well timbered watercourses around the north, east and south-east of Australia; seldom extends far inland. Sedentary. Other subspecies extend through New Guinea and neighbouring islands.

VOICE: In alarm it is a rapid **'ki-ki-ki-ki'**. Also a fairly slow **'swee-wit, swee-wit'**.

PREY: Birds, small mammals, reptiles and insects. At the nest we studied, birds constituted almost the entire prey recorded, the exception being one half-grown rabbit.

NEST: A large structure of sticks and twigs, lined with green leaves, usually on a horizontal branch, though sometimes against the main trunk of a tall tree, at a height of 10 - 30 metres or more above ground. Although typical nests have been described as shallow, the nest we studied was quite deep, about 350mm, and about 500mm in diameter. This nest had been used for many years.

EGGS: Two to four, usually three, 48 x 40mm. They are rounded ovals, without gloss, faint bluish-white, occasionally sparingly blotched with red-brown. The one clutch we recorded consisted of two eggs, almost unmarked, one egg having only a few smudges of brown. Egg-laying is usually from August to December. Assuming an incubation period of 35 days, the eggs we recorded would have been laid at the end of September. Chicks fledge in about 35 days.

145

INTERBREEDING OF GREY AND BROWN GOSHAWKS

In late September 1975 David phoned to say that he had a Brown Goshawk nesting near Orbost and felt sure her mate was a white one. His assumptions were based on the fact that he'd seen a White Goshawk carrying food, mostly rabbits, toward the nest area and he was certain there was only the one nest in that area. We never at any time doubted David's story, but to him we said we'd believe it when we saw the white one in the nest.

We were in the Orbost district later working on the White-bellied Sea-Eagle and while David did a stint on the latter we kept the Brown Goshawk under observation as she brooded her eggs, but the white male was not seen. It was decided we would work on these *accipiters* in four weeks time when there would be chicks in the nest.

Returning early in November, we set to work immediately to get the tower down a very steep, heavily timbered, bank. The dense undergrowth of briars and bracken was alive with leeches and they exacted a toll in blood. The work was back-breaking, and coming as it did immediately on top of a single-handed sixteen hundred kilometre non-stop drive from Chinchilla in Queensland, left one of the participants, to put it mildly and politely, physically exhausted.

The nest was at a height of twenty-two metres,

The tower at the interbreeding Goshawks' nest. The nest is twenty two metres above ground.

The female Brown Goshawk feeds her three chicks.

The clutch of three eggs.

but as we erected the tower higher up the bank we were able to get to nest level with only eighteen metres of tower. Between the three of us we kept the nest under surveillance for the next four days, during which the activities of the chicks and the brown female were photographed and filmed. At no time was the white male sighted near the nest, but he was often heard calling from a distance and the female would fly off and return a few moments later with a rabbit or part thereof. While David was in the hide the white male was seen carrying a rabbit toward the nest area, but the change over was not witnessed as it took place within the forest. We pulled out our gear on the 8th and returned home.

David phoned again on 22 November to say he had seen a Brown and a White Goshawk flying together in what he thought was a mating flight, although they didn't mate after alighting in a tree not far distant from him. A few days later he found a Brown Goshawk obviously on eggs. This nest was about a kilometre from the other, but in a far more accessible area.

We'd been using the tower we'd constructed for David on a Wedge-tailed Eagle's nest along the Murray River a few kilometres downstream from home. It was now standing in two metres of flood

Photographing the eggs of the interbreeding Goshawks

water so we tied floatation tanks to it before lowering it into the water and then towed it about two kilometres to dry land. We set it up at the second Goshawks' nest on 2 December. After a little gardening and photographing the eggs in close-up we watched from a distance to see the return of the female to the nest before leaving for home.

We left for Orbost again on 18 December and occupied the hide from 09:00 to 16:00 on the following day. The overnight trip of more than a thousand kilometres, when done alone, often left the photographer a bit weary, so he would play safe by tying himself in the hide. Our early hides had only a curtain for a rear door and falling asleep could have caused a rude, but brief awakening, followed by a permanent sleep. During those seven hours the female left the nest several times, but returned only with fresh sprigs of eucalypt leaves. The following morning the hide was entered at 06:00. The female commenced calling at 07:30, but got no response from the male. She left the nest three times during the next three hours and returned with green eucalypt leaves only. The chicks were very hungry and the female obviously most agitated as she called and peered about her. She would brood quietly for a few minutes, but the restless, hungry chicks gave her no peace of mind. At 11:00 she began calling most vigorously and left the nest, still calling. She could still be heard calling from a long way off, when, without a sound, the white male alighted on the nest. He peered in the direction of the hide for a few moments while the three tiny chicks solicited open-billed for food. He moved carefully over them and settled to brood. He didn't appear unduly worried by the sound of the running cine camera, nor that of the still camera shutter. Unfortunately the still camera was being reloaded and the cine stopped as the female alighted. The male flew off as she alighted and thus the chance of photographing them both together on the nest was missed. Although we kept the nest under observation for the next four days he didn't appear again.

As we did not see two White Goshawks in the area at any one time and a white one was seen to

147

The female Brown Goshawk brings green leaves to the nest.

The male White Goshawk pauses on the nest before brooding his recently hatched chicks.

travel between the two nest areas, we were led to the conclusion that he was mated with the two browns at the one time. It is also very unlikely that inter-breeding between another pair was taking place in the same area at the same time. At the first nest he was doing his normal job of providing for the brood, but it is doubtful if he did much hunting at all for his second mate and chicks. The brown female was forced to do her own hunting and this could have accounted for the disappearance of two of the chicks a few days later. There were plenty of predators in the area, such as Kookaburras, which would have taken the opportunity of preying on an unprotected nest, or they could have died of starvation and been removed from the nest by the female. She would, of necessity, be absent from the nest for extended periods while hunting, whereas at a nest where the male is doing his normal job the nest is only left unprotected for very short periods. The remaining chick was very slow fledging, probably due to being undernourished. We had our final look at it in January. Apart from looking rather emaciated it didn't appear much different to any other juvenile Brown Goshawk.

The almost fledged Goshawk chick.

Interbreeding in the wild of species in the genus *accipiter* is very rare. We know of no previous record of it between *A. novaehollandiae* and *A. fasciatus*. We have recorded it in the **Australian Bird Watcher,** whose editor, the late Roy Cooper, thought it of quite important significance, and in the interests of ornithologists had the transparencies reproduced in colour for the **A. B. W.** His main aim in doing so was to prove to all that the male bird was not an albino. We are indebted to Roy and his fellow trustees of the Ingram Trust in making funds available to offset the costs of the colour reproductions. The article has brought a big response from ornithologists and biologists in the U.S.A.

We worked the original nest again in the following season. The white male had been seen on the nest several times by David, but by the time we got there the chicks were past the stage where the male came in to brood them. While working the nest on that occasion the female swooped down to take a rabbit at the foot of the nest tree. It took her

a considerable time to kill it as they could be seen on occasions struggling in the undergrowth, but attempts to film the incident were ineffectual due to the undergrowth obscuring most of the action. After about half an hour she flew up to the nest with a portion of the rabbit. It was the first time we'd seen a raptor take prey close to its nest.

In 1977 David worked the pair again and got some good shots of the pair together on the nest. We were unable to get there at the time as we were making a film in the Strzelecki area with the A. B. C. We did work the nest later, but as in the previous season it was too late to get the male in. The pattern of behaviour was quite consistent from one season to the next.

He would do a very small portion of the egg incubation each day, but that was only while the female was off the nest feeding herself. For a short while after the chicks hatched he did a short stint each day, but thereafter his role appeared to be that of provider only.

In the following year we set up on the nest again. With the advantage of hindsight we had recruited a team of men to help carry the tower into position and stand it upright. It was only a few minutes work in contrast to the effort needed in earlier years. As the female was incubating eggs we had to be more cautious with our preparations. We didn't raise the tower above its basic height for the first day, but gradually got it to nest height over the next two days. The female was on five eggs, a very large clutch for a Brown and double the average for a White or Grey. On inspecting the nest on the morning of 27 September we found the white male incubating; it was a great relief to know he was still around and had us confident that at last we'd get them together on the nest. However it proved an unsuitable day to start work as it was cold and showery and we had no intention of risking the eggs to the elements for the sake of a day or two delay. On the following day the hide was raised a little above nest level without the female leaving the nest. This augured well for our eventual working of them as the hide was only four metres from her.

David did the first stint and the female sat tight till he was entering the hide, but she was back incubating only minutes after he settled in. The male, however, was more wary and did not do his usual short brooding session that day. At 06:30 next morning the male flew from the nest as we approached. This was greeted with mixed feelings; it was nice to know that he had accepted the hide, but would he come in again that day? There was only one way to find out and that was to sit it out and see. The pattern of behaviour that had emerged with David the previous year was his attendance at the nest each morning between 06:30 and 08:00 for from ten to twenty minutes while she presumably fed. On this occasion the female was back on the nest before the seeing-in party was out of sight. She sat quietly for three hours then left the nest for a few minutes. On her return she called a few times and left again. Moments later the male alighted and immediately settled to incubate the eggs. After forty minutes he apparently considered he'd done his share for the time being as he became restless and peered about him. He gave a few plaintive calls and soon afterwards the female returned. He was quick to give up his position to her, and the photographer had to move quickly to get two hasty shots of them together on the nest. David arrived moments later to see out a very jubilant photographer.

The hide was entered at 06:00 on 30 September and the female didn't bother to leave the nest. She left the nest at 08:50 and was immediately replaced by the male. She was back at 09:20 and there was barely time for one shot. They didn't co-operate with the photographer at all. At some landings she was directly behind him and thus partly hidden, at others he was on the move as she

alighted. Thus there was movement in the latter shots and not what we wanted in the former. It was very frustrating and time-consuming as there was seldom more than one changeover each day, but of course we kept on trying. David had a session on 1 October and got the male in about 08:00. On Monday we used the cine camera, but weren't optimistic about the results as it was cold and wet. Tuesday was a repetition of Monday, so we headed home Wednesday after seeing David in early that morning. We intended to get back after the chicks hatched, but were just too busy elsewhere. We left the tower for David to take down, but leave in the forest and so save on work in the following season.

According to David the five eggs hatched, but only three chicks survived. This was not surprising as the chicks tend to hatch progressively, thus giving the last little chance against the earlier

Back at the original nest, the male Goshawk comes in to incubate.

arrivals. In the interests of ornithology we felt that a chick should have been sent to the C.S.I.R.O. Wild Life Division to ascertain whether the chicks were infertile or otherwise. David didn't appear enthusiastic about the suggestion, so we didn't labour the point. However, a female juvenile was trapped and given to Mrs. Anne Troy of Melbourne, who devotes much of her time to the rehabilitation of sick and injured birds, particularly raptors. We photographed this bird in mid-1979, at which time it appeared to be developing into a typical grey phase Grey Goshawk.

In 1979 the white male moved to a new nest, a few kilometres from his previous nest, at a height of 38.5 metres (127 ft). While once again his mate was a Brown Goshawk we wondered if the new nest indicated he had taken a new mate. As far as we could see his bigamous behaviour was not repeated after the first season. To work this new nest would have required the addition of another section to our tower, but as we had obtained reasonable shots in previous years and were rather busy elsewhere, we decided not to try any further work on them for the time being.

David found the strange duo breeding again in 1980, once again in a new nest out of our reach as our tallest tower had been severely damaged at the Sea-eagle's nest, and the chicks had flown by the time it had been rebuilt.

By this time Anne Troy's captive female appeared to us and to most others who had no extensive knowledge of the species, to be still a typical Grey Goshawk. However, naturalist David Fleay who, with a lifetime of study of raptors, both in the wild and in captivity, is eminently more qualified to judge, considered there were subtle differences, characteristics of both parent species being detectable. The hybrid bird, in the meantime,

The female Goshawk returns to the nest and calls to the brooding male.

The female hybrid Goshawk, seven months after fledging.

appeared to have become attracted to another of Anne's charges - a white male goshawk - and he to her. Bernie Mace, who has studied *accipiters* for many years, undertook an attempt to mate the two birds in an elaborate aviary built especially for the purpose. The aviary, built largely by Bernie himself, with help from A. R. A. members, was ready early in 1981 and we were looking forward to learning more about the inter-relationships between members of the *accipiter* genus.

Unfortunately the hybrid bird met with a fatal accident soon after being given into the care of Bernie in 1981, so we still don't know whether or not it was capable of breeding.

Rear view of the hybrid Goshawk.

153

WHITE-BELLIED SEA-EAGLE
Haliaeetus leucogaster

A White-bellied Sea-Eagles' nest on a coastal cliff face in South Australia.

This species is our second largest raptor, with a wingspan of up to two metres. It is often known as the White-breasted Sea-Eagle, although White-bellied Sea-Eagle is its official title. It is fairly common along all coasts of Australia and extends well inland along coastal rivers. It is sedentary on some larger lakes and rivers of the Inland - one pair nest regularly on the Coongie Lakes in north-east South Australia. They are often found in the very tall Red Gums along the Murray River, and the Hattah Lakes - Kulkyne National Park is a popular site for the species.

As the Sea-Eagle adds a little more material to its nest at the beginning of each breeding season the nests become huge structures. We found one in the Kulkyne Forest which was at least four metres deep and contained possibly half a tonne of wood. In coastal regions the nests are often built on a ledge of a cliff-face or on a rocky out-crop. Aboreal nests are usually at least twenty to thirty metres above ground, although smaller trees are sometimes used. The species sometimes defends its nest against human intruders. More often, however, they will watch the intruder from a nearby tree, or circle above their nest.

Late in September 1975 we worked at a nest containing a single chick. The nest, in the main fork of a large eucalypt in the Orbost district of south-east Victoria, was at a height of twenty-two metres and was at that time the highest point from which we'd worked. Fish appeared to be their main food and we were intrigued by an aerial display performed by this pair on several occasions as we watched them circling the nest tree with a fish in their talons. They would release the fish and then dive to neatly take it again before it reached the ground. They never missed retaking it at their first attempt, demonstrating their expertise in taking prey in the air or sea.

Although we recorded only fish as prey at this nest the Sea-Eagle will also take other aquatic and marine creatures, such as tortoises and sea snakes, and small mammals, such as rabbits. They will also feed on carrion. In 1978, beneath a nest on the lower reaches of the McArthur River in the Northern Territory, we found the ground strewn with many tortoise shells, some up to 250mm in length, while evidence of fish as prey was quite meagre.

The quality of our work at the nest in the Orbost district was disappointing and we always intended to work the species again. The opportunity came in September 1980. Two A. R. A. members, David Baker-Gabb and Max Arney, had banded two chicks in a nest in the Hattah-Kulkyne National Park, but that didn't deter us from working them after seeking and receiving permission from the park ranger. The nest was at a height of nineteen metres and needed a little gardening to make it suitable for filming. This had to be done while standing in the nest, and to get into it one's face had to be brought very close to the two chicks, which by that time were starting to fledge and would have made very formidable opponents if they had been that way inclined. Fortunately they remained passive. One was considerably larger than the other and when we started to work them we soon realized the biggest contributing factor to their difference in size was a dietary one.

The hide was entered early in the morning to ensure that the two chicks weren't fed before we entered, and thus avoid a long wait for the next feed, which might be ten or more hours away. It was obvious from the state of their crops that they hadn't been fed that morning. The smaller one's crop was empty and the chick was already making weak croaking calls for food. The parent birds were apparently hunting or they would have been circling the nest-tree as we approached the hide. The larger chick still appeared to have a little food in its crop, but it also called occasionally.

For four hours they called intermittantly, but it was at 11:00 that their vigorous calls and obvious excitement heralded the approach of the female. She alighted with a small carp in her talons and, as she appeared apprehensive of the now occupied hide, no shots were taken to try and allay her fears

The White-bellied Sea-Eagle in flight.

A typical clutch of Sea-Eagle eggs.

Our tower at the Sea-Eagles' nest in the Hattah-Kulkyne National Park.

of it. The small fish was taken by the larger chick and any movement by the smaller one to lift its head, or even glance toward the fish, brought a response in the form of a two syllable protesting squawk, or a peck, or both. The smaller one was quite cowed. The protesting squawk was enough to make it quickly lower its head if it had indeed raised it at all. The female left the nest and the larger chick managed to tear apart the carp and fed itself. It was probably the first time it had been left to do that. The smaller chick still croaked out calls for food and around 12:30 the female arrived with substantial prey, a water bird of some species, not possible to identify, being almost bare of feathers. For the next hour the female steadily fed the larger chick. Anything that could be interpreted as a move to proffer a morsel to the smaller one brought a protesting squawk from the other. The adult seemed to always react to the protests and the larger got the proffering. Even when its crop was obviously full and the skin around it appeared drum tight it continued to protest if the smaller looked like getting anything. The chicks of several raptor species establish a pecking order and in the case of some, death from starvation does occur. Mostly, however, once the dominant chick is reasonably sated the next in line feeds. Here, however, was an exhibition of the 'dog in the manger' attitude that was unsurpassed at any nest we've worked. It was, from the *Homo sapien*'s point of view, quite disgusting, but

The female Sea-Eagle with her two chicks: the smaller chick cowers with its head down.

156

never-the-less the way of nature, survival of the fittest.

The effects of so much food were gradually causing a lethargy in the larger chick and the smaller would sneak a morsel without lifting its head. It almost appeared as if there was a conspiracy between it and the adult as she too, kept her head low, only turning her bill with the morsel toward the chick. By 14:00 nothing remained of the prey, but the small chick was far from sated. The adult flew off, it is hoped, to get further prey for the soliciting chick. The photographer also left, very cramped and hungry, after seven and a half hours in that tiny hide twenty metres above ground and seven from the nest.

Five days later we worked them again. The female had by that time got used to the hide and accepted it as empty provided the seeing-in person climbed to the hide door before departing. She obviously couldn't count. It appeared to us that the chicks weren't being brooded at night as we never saw an adult bird leave the nest even when we got there before daybreak. Hunting must have started by sunrise on this occasion as the hide was entered before sunrise and shortly after the sun had risen the female alighted with a fish weighing at least half a kilogram. The larger chick ate 90% of it, the balance being eaten by the adult when a piece seemed unsuitable for the chick. The smaller chick got nothing to eat, but it did get

The surviving Sea-Eagle chick exercises its wings.

157

some nasty pecking when the other attacked it viciously. The victim cowered with its head between its legs, but the other still managed to reach under and pull feathers from its head and neck. The adult bird took no notice of these attacks and just waited with a morsel of fish for the attacker to take when it had finished venting its spleen.

We returned late in October from working in northern Queensland to find our tower on this species a twisted heap of metal. Guy ropes had broken during a wild storm. There was only the one chick in the nest which didn't surprise us as the prognosis for the smaller one when last observed was not good. It had no doubt succumbed under the repeated attacks on it and died of malnutrition or injury or a combination of both. Nature can appear very cruel.

On 1 November 1980 another tower was set up on the nest in an endeavour to get a shot of the now fully-fledged juvenile and an adult together on the nest to point up the contrast in plumage. The female was very wary and we didn't get her to remain in the nest till the third attempt. The chick was able to feed itself and did so, but on the third day she remained in the nest after the chick had shown its independence by ignoring a proffering from her and grasped the waterbird, probably a Eurasian Coot, that she had brought in, and fed itself. She found the remains of a partly eaten fish in the cup of the nest from which she managed to get a few pickings. She occasionally proffered a morsel to her offspring but was, for the most part, ignored. It was now ten weeks old and capable of flying if put to the test. When exercising its wings it often became airborne, drifting fore, aft and sideways of the nest, but always managed to alight again on it. Unless unduly provoked by human intruders at nest level it would probably remain in the nest for another week.

The dark brown plumage of the fully fledged Sea-Eagle chick contrasts sharply with that of its parent.

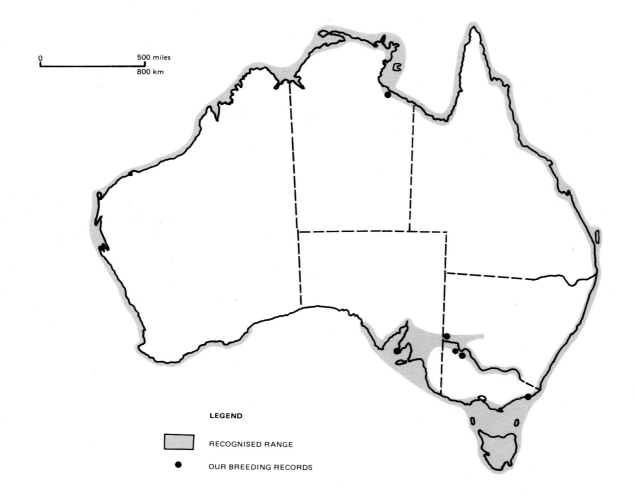

LEGEND

▭ RECOGNISED RANGE

● OUR BREEDING RECORDS

WHITE-BELLIED SEA-EAGLE
Haliaeetus leucogaster

hals - sea (Gk); *aetos* - eagle (Gk); *leucos* - white (Gk); *gaster* - belly (Gk).

OTHER NAMES: White-breasted Sea-Eagle; White-bellied Fish-hawk.

LENGTH: 710 - 890mm.

WINGSPAN: 1750 - 2000mm.

DISTRIBUTION: Moderately common and sedentary around the Australian coast and nearby islands; sometimes extends well inland along coastal rivers. Also sedentary on some larger lakes and rivers of the inland. Also from New Guinea and neighbouring islands to Indonesia, South China and India.

VOICE: Loud, somewhat goose-like **cank-cank-cank'**. Also a more rapid cackle.

PREY: Fish, sea-snakes, tortoises, water birds and mammals such as rabbits. Prey brought to nests under our observation consisted of fish and water birds. However, beneath a nest we found on the lower reaches of the McArthur River in the Northern Territory there were many tortoise shells up to 250mm in length. Sea-Eagles will also feed on carrion.

NEST: A huge structure of sticks, lined with leaves, placed in a main fork of a tree, on a cliff ledge, on rocks or on the ground. We saw nests, both aboreal and terrestrial, over three metres deep, but more often they are 1.5 to 2 metres deep. Usually aboreal nests are 20 - 30 metres or more above ground, though smaller trees are occasionally used.

EGGS: Usually two, rarely one or three, 71 x 53mm. They are white, rather long ovals. Of six clutches we found: 5 were of two eggs and 1 of one egg. Egg-laying is from May to September, but all eggs found by us have been laid in July. Although we have been unable to accurately establish incubation period, we would assume it to be about six weeks. Chicks fledge in about ten weeks.

WEDGE-TAILED EAGLE
Aquila audax

Perhaps no other raptor is so well known in Australia as the Wedge-tailed Eagle. It is our largest raptor, with a wingspan of two to two and a half metres in most cases. The Latin name for the bird *Aquila audax*, meaning 'bold eagle', befits the birds appearance, but it is really a very wary bird, particularly in more closely settled areas. It will not normally defend its nest against human intruders, but will leave furtively and stay well away.

Until recently this species was slaughtered in large numbers throughout the nation. Some states paid bounties for its destruction. In Western Australia, in a forty one year period to 1968, a total of 147,237 five shilling bounties were paid by that state. Bounties ceased in November 1968.[12]

In one year almost 10,000 ten shilling bounties were paid in Queensland. One old-timer related to us how he and many others earned pocket money as boys collecting the bounty in the latter state. At that time he said the bounty was two shillings (twenty cents) for birds, a pair of claws, and six pence (five cents) each for eggs. On finding a nest with eggs they had to decide whether to take the eggs or leave them and perhaps collect the higher bounty for chicks. The decision was not quite as simple as one might deem it to be, as there was always a chance that another lad might find the nest and decide that an egg in the hand was worth more than two in the bush.

Fortunately the species has survived that misguided onslaught and is being recognized for the good it does, especially in helping to control the rabbit pest. We've found scores of rabbit skulls on the ground beneath this species' nests, mute testimony to their predation on rabbits. As it is also a carrion feeder it must play a part in bush hygiene by removing the source of much fly infestation.

A typical Wedge-tailed Eagle nest and eggs.

A fully fledged Eagle chick exercises its wings.

The nest is a huge structure of large, dead sticks, usually placed in a tree, but on rare occasions, on a low bush or on rocks. Each pair of birds often has a second or third nest within their territory from which to choose each season. There appears no set pattern as to which nest may be chosen. They may use the same nest for a number of seasons or change each one, but they do appear to remain territorial. The nests are usually well lined with eucalypt leaves during the breeding season, being added to several times a day while the chicks are in the nest. There is a suggestion that the fresh eucalypt leaves help to keep the nest free of vermin or to increase humidity.

Nests vary greatly as to height above ground. In the inland areas where trees are often small and stunted, we've come across nests that we could peer into while standing on the ground, while along the Murray River we found a pair nesting at 30 metres.

Two eggs usually form a clutch, but three are occasionally found. Both eggs usually hatch, but in many cases only one chick will survive as the stronger one may kill the other. From our observations we don't accept this Cain and Abel drama as inevitable. In some areas and under certain conditions more twins are reared than lone chicks. It probably has a lot to do with the availability of food at the nest and the adults' presence on the nest during the critical period when the killing is most likely to take place. The adult can be off the nest for two main reasons: the collection of prey and as a result of human intrusion. Aggression can't take place while they're being brooded and they are normally brooded fairly consistently for the first two or three weeks. Therefore if the birds are not disturbed and prey is plentiful most nests will rear two chicks.

We made our first attempt to photograph Wedge-tails in 1973, before we had consciously embarked on our project, and before we had built our first tower.

Early in October, about thirty kilometres from home, we found a nest containing one well fledged chick. The nest was about fifteen metres up a large Red Gum on the bank of a small creek. Normally the creek bed would have been dry, but with the nearby Murray River in flood, the nest tree was surrounded by flood-waters. Undaunted, we decided we would build a hide in the nest tree, on a horizontal branch overlooking the nest. We set about the task next day.

Due to the floodwaters, we were unable to reach the vicinity of the nest tree by vehicle, so the last five kilometres were travelled in our small outboard powered dinghy. This we loaded with a six metre ladder, assorted pieces of timber and the various tools we might need to build a platform in the tree.

Within two hours we had a reasonably level and secure platform, about five metres from the nest. This we piled high with eucalypt branches and, with a rising sense of excitement, we made our way home.

By sunrise next morning we were once more boating through the muddy brown waters towards the nest tree. In the boat this time we had our hide and cameras. Today, if all went well, we would get our first raptor photographs.

As we came in sight of the nest, the female Eagle left the nest tree, and with deep, powerful beats of her huge wings, soon disappeared around a bend in the creek. We silently wondered how long it would be before she returned.

It was a relatively simple task to haul the hide up onto the platform and camouflage it with branches. By 09:00 everything was ready. The hide was to be occupied until 17:00, by which time we felt sure we would have some results.

As the sound of the outboard motor faded away into the distance, the long vigil began.

The chick spent the morning quietly preening itself or dozing in the nest. Early in the afternoon it began scanning the sky and calling, but after a few minutes it settled down once again to preening and dozing.

The occupant of the hide, his initial excitement somewhat dampened as the hours dragged by, was himself almost dozing when the first faint sounds of the outboard motor signalled the end of his self-imposed confinement. He was down to the foot of the tree by the time the boat arrived.

Despite our disappointment, we began earlier next day and were at the nest tree soon after sunrise. Once again the morning passed uneventfully. Occasionally the chick pecked half-heartedly at a rabbit carcass in the bottom of the nest.

Suddenly, at about 14:00, the chick began calling excitedly. Moments later the female Eagle landed on the back of the nest, a sprig of leaves clutched in her massive talons. Scarcely daring to breathe, the photographer stared in awe into the fierce eye of the Eagle, seemingly only centimetres from the end of the lens. Cautiously he adjusted the camera and pressed the shutter

Our first Eagle photograph.

release. We had our first Eagle shot.

The Eagle only stayed long enough for a second shot before hurriedly departing. Surprisingly, less than an hour passed before she returned, this time with a rabbit. However she stayed only long enough for two more photographs before flying to a branch higher up the tree. There she spent the rest of the afternoon, being constantly harassed by a Willie Wagtail, until the sound of the approaching boat put her to flight.

Though a few days later we undertook another eleven hour session in the hide we were unable to obtain any further photographs.

In 1974 we worked a nest along the Darling River in south-west N. S. W. We kept the tower well back from the nest and soon had both adult birds coming regularly to the nest with rabbits for their two chicks.

At their nest in south-west New South Wales both adult Eagles tend their two chicks.

The female Eagle feeds the part fledged chicks.

The following year we found an Eagle incubating two eggs not far from the 1973 nest. We set up the tower soon after the eggs hatched. From the outset it was apparent that the smaller chick would be lucky to survive. It had been severely pecked about the head, so we weren't surprised to find it dead in the nest on our next visit.

Once again rising floodwaters made things a little difficult for us. We soon had to use the boat to reach the tower, and by the time we had finished filming for the season, it was standing in two metres of water.

By the time the surviving chick has fledged, flood-waters surround the nest tree.

The Eagles' nest in a River Gum near the Murray River.

It has been said that birds kill only for their requirements or that of their brood. However, with rabbit in plague proportions along the Strzelecki Track in 1976 we found a nest containing two recently hatched chicks and laden with no fewer than twenty uneaten rabbit carcasses, most of them appearing to have been killed within the previous two or three days. It appeared that the adult Eagles may have gone into some sort of 'overkill reaction' after their eggs hatched. We found that while chicks were being reared nests are usually kept fairly clean, most prey remains being removed regularly. However, occasionally, some less 'house-proud' birds allow uneaten prey to accumulate over a period of weeks, resulting in a situation offensive to both eyes and nose.

With such ideal conditions prevailing in 1976, all nests we inspected reared two chicks. We set up our tower about five metres from one nest along the Strzelecki and the female didn't leave the nest during that operation. Nor did she move when the tower was climbed and the hide entered. That was in complete contrast to their behaviour at any nest we'd worked previously.

It was intriguing to observe and record the solicitude of that female for her two chicks. They were, to put it simply, pampered; she fed them in bed!

An Eagle chick lies on the edge of its nest, almost dead from repeated attacks by its sibling.

Two newly hatched Eagle chicks are surrounded by more than twenty rabbit carcasses.

The 'tame' Eagle with her two part fledged chicks.

When they had obviously had enough to eat and were dozing in the cup of the nest she would sometimes walk over to tempt them with another morsel of rabbit flesh, which one might or might not take. After proffering it several times to the one without it being taken she would repeat the performance with the other and if not taken, eat it herself. When sunlight fell on the nest she would stand with wings draped to shade them till the shadows of the foliage again fell on them. She brought in eucalypt leaves many times a day to spread on the huge nest. So big was the nest and so deep with leaves that it would have made a comfortable bed for one or two people.

The male seldom came to the nest, but did all the hunting and the female would fly out and collect the rabbit from him. It was the only prey fed to those chicks and it was roughly skinned on the nest before feeding them tiny morsels. Anything she considered too large for them to swallow easily she would try and grasp in a talon or failing that would place a talon over it to anchor it while attempting to tear smaller pieces from it. If not successful she would eat it herself. Her movements on the nest were very slow and careful in her care not to harm a chick. One never fails to feel amazement and wonder in that all that beautiful exhibition of motherhood in the animal (bird) is inherent and not reasoned or learned as with *Homo sapiens*.

The chicks are a downy white for the first two or three weeks, after which the wing and tail primaries start to appear as a dark line amid the down. They are ugly ducklings from about five to eight weeks, being a patchwork of down and feathers, the last few weeks seeing them emerge as elegant as their parents. Fully-fledged the juveniles are generally much lighter in colour than their parents unless the latter happen to be relatively young. With each moult they tend to become marginally darker. Old birds are almost black having only a narrow band of bronze-like feathers along the upper surface of the wings. The older and darker they become the more majestic they appear.

The fledging period in one case was seventy-five days, but this can vary a lot as some juveniles are more reluctant to make that first flight than others. We've noticed that a fully-fledged youngster may be flying from nest to branches and back for some days before making its first flight clear of the nest-tree. Incubation has not been easy to ascertain precisely, but we estimate it to be around forty-five days.

In 1977 all nests that we inspected along the Strzelecki again reared two chicks and a nest was found by rabbiters that contained three, one of which they took and reared as a pet. We were disappointed in not being able to record such a rare event on film before the chick was taken.

One nest that we worked that year had three adults in attendance although only two appeared on the nest and both fed the chicks, while a third brought prey to within about a hundred metres of the nest. It appeared that two females were feeding the chicks as it is doubtful if a second male would be welcome.

Breeding began early in 1979, some birds having chicks in June before others had yet laid. June is the approximate time that breeding starts normally, but seasonal conditions were far from normal with rainfall double the annual average in January alone. A nest containing three eggs was photographed and again forty-six days later when there were three chicks, the smallest of which was no more than two days old. It had a peck mark on its head which, coupled with its small size in relation to the others, didn't make the prognosis a cheerful one. We weren't surprised therefore when we found the brood reduced to two a few days later. Indeed, a subsequent visit some weeks later revealed only one surviving chick, well fledged and healthy.

On 20 July 1979, while searching for the Grey Falcon, we found the 'tame' Wedge-tail that we'd worked in 1976 was back in that nest after an absence of two seasons. She had no doubt nested in an alternative nest within her territory in the interval. She appeared even more tolerant toward us while incubating than she was with chicks almost three years earlier. In turn we climbed to the nest and talked quietly to her while she sat tight and answered with weak protests, perhaps best described as low wheezing squawks. As one leaned over the nest toward her she stood up with wings outstretched, and shots of her with the eggs at her feet, were taken, using a wide-angle lens. If we leaned too close to her she would slap us with her wings, but would not give ground. We had no doubts that she would strike with bill or talons had we attempted to touch her eggs. Needless to say we didn't try, and besides, we didn't want to put the eggs at risk of being broken.

In the years since we'd last worked her there had been a distinct change in her plumage. She was now a darker bird with less gold braid. It is our intention to photograph her each season, if we can locate her nest, to record plumage change.

The Wedge-tailed Eagle in flight.

At one of the many nests containing two chicks in 1977 the female Eagle brings leaves for nest lining.

A rare clutch of three eggs.

The three newly hatched chicks seek shade. Only one chick survived to fledge.

The 'tame' Eagle in 1979 - her plumage is noticeably darker.

From her nest the Eagle surveys the surrounding Strzelecki flood plain. Over the preceding three years she has lost much of the golden buff colouring from her upper back and wings.

The female Eagle advances with wings beating and bill agape.

There were another three pairs of Wedge-tails nesting within two kilometres of her nest, one of them only a kilometre distant. Further along the creek we found two pairs nesting only seven hundred metres apart. It appeared that some of these eagles were nesting within another's territory and using the latter's alternative nests. Such close nesting of the species had never been observed by us before although most of the nests were known to us, but seldom used. We would speculate that this close nesting was only tolerated due to the abundance of prey. Rabbits were in plague numbers.

On 8 August we set up at the 'tame' one's nest to film her reaction to an intruder. As we knew of no other instance of this species defending its nest in any shape or form we thought it advisable to have proof of our claims. She acted more dramatically than we had dared hope. As the intruder moved cautiously along the bough toward her nest she rose and advanced to the nest edge to tower over him with outstretched wings and open bill. In the face of this most daunting stance he raised an arm to protect his face and eyes, and edged a few centimetres closer. So fast did her talons flash out that neither he nor the cameraman saw what struck him. He withdrew with a hand

dripping blood. She glared after him for a moment before turning and settling to brood her eggs, while a very relieved cameraman then stopped filming. We removed the tower as it was needed elsewhere.

We set up at the same nest again on 28 August and from her posture we suspected she was brooding chicks. She appeared to be expecting the male in as she scanned the sky above and all around. Occasionally she moved around a little, but never raised herself sufficiently to reveal her brood, but for most of the time just sat quietly the only movement being the nictitating membrane over her eye.* For lack of something better to do the number of times that membrane was activated was counted; it averaged thirty times per minute. Sometimes something probably attracted her attention as she stared for several seconds with it withdrawn, while at other times movement was once per second.

---

* *A thin membrane by which the process of winking is performed in certain animals, and which covers and protects the eyes from dust or from too much light. It is found in birds and reptiles and is sometimes referred to as an extra eyelid.*

166

The female Eagle alights with leaves which she has torn from the nest tree.

She alights with a rabbit . . . .

. . . . then feeds her two day old chicks.

After three rather boring hours for the camera-man she stood up to reveal two tiny chicks, probably only a day or two old. She flew off and in a few minutes returned with a sprig of eucalypt leaves which she just dropped on the nest before selecting from bits of rabbit carcass scattered around it. She eventually found a piece from which she managed to extract a few morsels for each chick. The long wait was well rewarded by the beautiful sight of such a fierce and powerful looking bird behaving so gently and delicately as she fed her tiny chicks. We had always found Wedge-tails fascinating but we found this bird particularly so.

She was a unique bird.

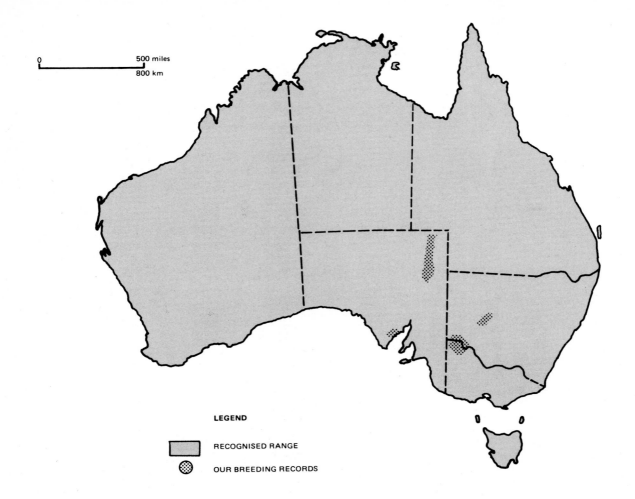

WEDGE-TAILED EAGLE *Aquila audax*

aquila - eagle (L); *audax* - bold (L)

OTHER NAMES: Eaglehawk.

LENGTH: Female approximately 1000mm. Male approximately 900mm.

WINGSPAN: 2000 - 2500mm.

DISTRIBUTION: Uncommon to common throughout Australia in a range of habitat, from mountain forest to wooded grasslands and open plains. Generally sedentary, but may become nomadic in times of drought.

VOICE: Most calls seem to be rather feeble, wheezing whistles and screams. In alarm it is usually a long wheezing scream, at other times a double 'pee-ya, pee-ya'.

PREY: Rabbits and other small mammals, particularly small kangaroos and wallabies, birds and reptiles. Carrion also figures largely in their diet. Undoubtedly rabbits are their main prey item where they occur. At all nests we worked rabbits were almost exclusively the prey recorded, the only other prey being a large bearded lizard.

NEST: A very large platform of sticks, copiously lined with green leaves, placed in a main fork of a tree, or occasionally on a rock-ledge or on the ground or low bush. We saw an active nest only about 700mm in diameter, but often they are 1 to 2.5 metres in diameter and over a metre deep. The centre cup is usually 300 - 400mm in diameter. When arboreal, height above ground varies widely depending on availability of suitable trees. We saw nests from 1.6 metres to 30 metres above ground. A pair of birds usually have one or two alternative nests in their territory and these may be used on an irregular basis.

EGGS: Usually two, occasionally only one and rarely three, 73 x 58mm. We have recorded about thirty clutches, all except three consisting of two eggs, there being two sets of three and one single egg. They are coarse rounded ovals, dull white or buff, spotted and blotched with purplish-brown and red-brown. One egg is always less marked, sometimes with only faint underlying purplish markings. Eggs are usually laid from June to October, chiefly July and August. Our records show egg-laying from early May to September. We estimate incubation at about 45 days and fledging 70 to 75 days.

AUSTRALIAN LITTLE EAGLE
Hieraaetus morphnoides

While the Australian Little Eagle, or Little Eagle as it is more commonly known, is found throughout Australia, we have found it most common in inland areas, where it is often seen soaring over woodlands and timbered water-courses.

There is a wide range in plumage colour. The dark phase, which is rather rare, could be confused with a Black-breasted Buzzard, but lacks the upswept, distinctly "bullseyed" wings. The more common pale phase is somewhat similar to the Whistling Kite in size and general colour, but can be distinguished in flight by its shorter, broader tail and less-fingered wings which are carried flat - with occasionally a slight upturn at the tips - in contrast to the kite's which are distinctly fingered and bow downwards. When perched the rather heavy, feathered legs of the Little Eagle contrast with the lighter bare legs of the kite. The scientific name derives from the Greek words *hierax* meaning 'hawk', *aetos* meaning 'eagle', and *morphnos* for 'form', suggesting the Little Eagle has 'the form of an eagle-hawk'.

Although some Little Eagles are nomadic, many appear sedentary, often using the same nest for a number of years in succession. The nest is a large stick structure, lined with green leaves and usually placed high in the fork of a tree. One or two eggs form the clutch - there is some evidence to suggest that clutches of two have become more common since the proliferation of the rabbit - but seldom more than one chick is reared. They have proved delightful birds to work as they take little notice of the hide after the first day and go about their nest activities without apparent inhibition.

The Little Eagle in flight.

Little Eagle eggs: one or two eggs form a clutch.

In 1974, in the Millewa wheat country of north-western Victoria, we worked a pair nesting in a belar tree at a height of twelve metres. The chick was only two days old and the female was back brooding it within a couple of minutes of the seeing-in party leaving. She was probably the

The female Little Eagle at her nest in the Millewa region of north-west Victoria. In this area the Little Eagles show a distinct preference for Belar trees as nesting sites.

The female Little Eagle alights at the nest.

Both adults shared nest duties at this nest.

equal of that Wedge-tailed Eagle we worked on the Strzelecki in later years, in her maternal care for the chick. Only the tiniest morsels were proffered, anything larger being eaten by herself. Her movements were equally as careful as her feeding routine. Advancing to brood the chick her feet were moved slowly and tentatively. The male brought all prey to the nest while the chick was being brooded. On occasions she left the nest - ostensibly to feed herself - and the male would come in to brood the chick after she left. His movements were also slow and careful, like action depicted in slow-motion. Prey was mostly rabbits with an occasional small bird, such as a Pipit.

The nest was used again in the following year and one chick reared. Behaviour and routine seldom varied, green leaves were brought in daily. This was the first chore of the day by the male in the early part of the fledging but was by either or both birds later.

In 1977 we located another nest in the Millewa, again in a belar tree and only a few hundred metres from a Peregrine Falcon's eyrie. As is usual for the species, whenever we approached or climbed the nest-tree, the Little Eagles left the nest area completely. However, we seldom reached the nest-tree without the Falcons raising a protest.

In October the following year this nest, and another some kilometres away, were found to contain two eggs in each, so we kept them under observation in an endeavour to ascertain the fate of a second chick, should it hatch. There were two eggs in each nest on 31 October, and on 3 November, each had a chick and an egg. A week later one had two lively chicks and the other a week-old chick but no egg. Shell in the outer edge of the nest indicated it had hatched but the chick had presumably died or been killed and the body removed.

The female Little Eagle with her ten day old chick. At this nest, the male did not participate in nest duties.

The pair of Little Eagles at their nest. The smaller male is an extremely pale bird.

We kept a close watch on the nest with two chicks. There appeared to be three or four days difference in their ages but there was little if any aggression between them. Anything that could have been construed as aggression was by the smaller toward the other. The female fed them simultaneously and both chicks and the adult were a delight to watch. Prey was entirely rabbit which was caught and brought in by the male to the nest. The male only stayed long enough to release the prey. This behaviour was at variance to that of those worked in 1974 and 1975 when the male often did some brooding and occasionally fed the chick. She did no hunting even when the chicks were well advanced and were no longer brooded during the day. She seldom left her perch on a nearby limb of the nest-tree, except when apparently off feeding herself. The moment the male alighted on the nest or beside it she would alight chattering beside him, take the prey in her talons and urge him to be gone if he hesitated in leaving. He obviously wasn't welcome to stay as her chatter didn't cease till he flew. His stay was seldom more than three or four seconds. On occasions the rabbit arrived intact and she plucked off most of the fur before feeding the chicks. She also carried the paunch and intestines some distance from the nest to dump them. On very hot days she stayed on the nest to shade the chicks.

On two occasions White-winged Choughs came around to tease her. They would sometimes get within a metre of her, chattering, but on the alert for a feint from her and she usually obliged with a quick duck of her head as if to take off at them. They would screech and spring to high branches

172

The female Little Eagle feeds her two tiny chicks.

The two fully fledged Little Eagle chicks.

The Little Eagle shows a distinct crest as it perches with its back to the breeze.

A pair of Little Eagles tend their chick in their nest along the Strzelecki Creek.

before starting the game again. When she appeared to tire of the game and just ignored them, they usually went their noisy way, hoping no doubt to find some other recipient for their attentions.

We followed the chicks' progress till they were 38 to 42 days old, at which time we removed the tower as we required it in Queensland. A fortnight later, estimating that the chicks would be nearly fledged, the tree was climbed in order to photograph them. However, the appearance of a human head over the edge of the nest caused both chicks to attempt to fly, but they landed clumsily on lower branches of the tree. The next half hour was spent retrieving them from their precarious perches and returning them to their nest where they were finally photographed. They were 52 to 56 days old and were much darker than their parents, especially on the breast, and very similar to dark phase birds.

In 1979 we briefly worked a pair nesting in a Coolibah on the Strzelecki Creek. As with the pair

we worked in 1974 and 1975 the male took an active role in brooding and feeding the three day old chick. The male bird was darker on the breast than his mate, though by no means could he be described as a dark phase bird - more an intermediate phase.

We had seen only a few birds that were truly dark, and had been unable to find one nesting until October 1980 when we set up our tower at a nest - once again in the Millewa, which contained one healthy week-old chick and an unhatched egg, and was tended by a very dark female bird, the darkest we had seen. We left the tower set up for about a fortnight before attempting to do any work at the nest as we were busy elsewhere. On the evening prior to our planned working, the flash heads and various gear was put in the hide ready for a start on the morrow. During the night there was a wind storm which wrecked our plans. The chick was found dead on the ground not far from the nest-tree next morning.

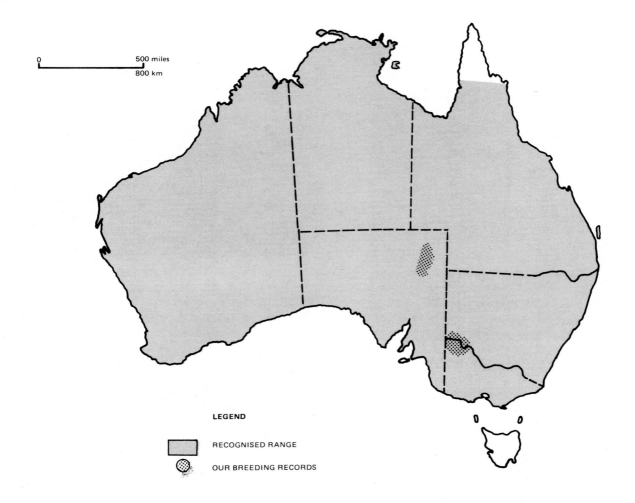

LEGEND

RECOGNISED RANGE

OUR BREEDING RECORDS

AUSTRALIAN LITTLE EAGLE
Hieraaetus morphnoides

hierax - hawk (Gk); *aetos* - eagle (Gk); *morphnos* - kind of eagle (Gk); *eidos* - form, or like (Gk).

OTHER NAME: Little Eagle.

LENGTH: 450 - 550mm. Female larger than male.

WINGSPAN: 1100 - 1200mm.

DISTRIBUTION: Uncommon to moderately common in wooded areas throughout Australia (subspecies *H. m. morphnoides*). Sedentary. Also in New Guinea.

VOICE: Usually a short double or treble whistle, the second and third notes being slightly shorter and lower pitched.

PREY: Rabbits and other small mammals, reptiles and birds. Carrion is also occasionally eaten. Rabbits were the predominant prey observed by us, but birds also figured largely at some nests. Pipits and quail were species identified.

NEST: A large stick structure, lined with green leaves, usually placed high in a fork of a tree, 10 - 45 metres above ground.

EGGS: One or two, 55 x 44mm, coarse bluish-white, rounded ovals, unmarked or sparsely streaked and blotched with red-brown. We found clutches of two eggs more common than single eggs: (9:3). Eggs are laid from August to October, our records indicating from early August to late September. The incubation period is not less than 32 days, with fledging about 50 days.

174

PACIFIC BAZA
Avideda subcristata

The beautiful, quiet and generally unobstrusive raptor, often called the Crested Hawk, can be found from the north-west of Western Australia, across the top and down the eastern coast of Queensland as far south as the north-east coast of New South Wales. The north-western subspecies is slightly smaller than the eastern form. The Baza is more common generally in coastal regions, but we have photographed it up to three hundred kilometres inland, which appears to be as far from the coast as it extends.

They usually nest at heights between fifteen and twenty-five metres. All nests that we have found to date have been in Eucalyptus species or Apple Box.

The nest is a flimsy structure of leaves and twigs. They are often blown away soon after the young have left the nest, and we suspect sometimes before, but we were shown a nest not far from Maryborough, Queensland, which, it was claimed, had been used for several seasons in succession.

We worked a nest in January 1975 in a eucalypt on the bank of the Mary River on the outskirts of the city of Maryborough, Queensland. The nest was at a height of sixteen metres and close to a busy highway and railway, but the noisy traffic on both did not appear to bother this quiet bird. We had first to erect the tower close to the nest to do

some gardening and while engaged in that one of the two chicks fluttered out of the nest. We had to desist from any further gardening work or we may have had nothing to photograph in the nest. As a consequence our shots weren't as good as we would have liked.

We caught the chick later and returned it to the nest. The catching of it was by far the easiest part, the placing of it in the nest with the use of a hat fastened on the end of a five metre length of wooden cover strip, while working from the hide sixteen metres up and five metres from the nest, was the hard bit. For some time the juvenile could not be induced to vacate the hat for the nest. Any sudden movement of the hat while near the nest might cause one or both to flutter to earth and return the situation back to square one or beyond. When aching arms were approaching their limits of endurance the obstinate juvenile stepped calmly from hat to nest.

On Boxing Day 1975 we left for another attempt to get some further shots of the species nesting near Wandoan, Queensland. Peter Slater had had the nest under observation for some time and was thus able to supply data not otherwise available to us. We are indebted to Peter and his wife Pat for their hospitality and the chance to work the species again.

The female Baza watches over her small chicks.

The nest was only twelve metres up in a eucalypt beside a small creek. It was the lowest nest we'd seen and we worked it finally from a distance of two and a half metres. The two chicks were six days old when we moved the tower to its final position, from where we got some excellent photographs and film.

The chicks were being fed almost hourly on frogs, grubs and insects, especially phasmids. The size and colour combinations amazed us. Many of them are seldom seen by man as they live in the tree-tops where their stick and leaf-like shapes and colour blend with the foliage. They are also known as stick insects and it speaks volumes for this bird's vision in detecting them. Prey was always carried in the bill.

Both adults were feeding the chicks till the tower was erected, but thereafter only the female came to the nest. She became very tame and ignored us as we entered or vacated the hide. We learned later of a pair elsewhere being worked without a hide, the photographers experiencing much the same reaction as ourselves.

Incubation for this brood was thirty-three days and we estimate fledging would be about the same. Unfortunately one chick was blown, knocked or fell from the nest a week or so after our departure. Due to gales and the flimsy structure of Bazas' nests, we suspect that mortality is high during the fledging period.

On 14 December 1978, following a phone call from Chris Cameron, we left for 'Rockwood', near Chinchilla, to work them nesting on the property close to where we'd worked the Square-tailed Kites in 1976. Approaching the nest area we were perturbed to see many fallen tree branches and up-rooted trees along the highway. A very recent storm had gone through the area and we were pessimistic about the chances of the chicks being alive. However, the nest was still intact with two chicks in it; the remains of a third, smothered in ants was at the foot of the nest-tree. Whether it had been blown from the nest or just fallen, it tends to reinforce our earlier suggestion that many of the species die in that manner. Even at a height of twenty-three metres the nest-tree was some-

The male Bazza pauses beside the nest with an insect for the chicks.

177

what protected by the surrounding forest of pine trees. Thunder storms accompanied by gale force winds are quite common in that area at that time of the year.

We hacked a path through the trees enabling us to drive to the nest-tree, an Apple Box. After two hours we had the tower erected close enough to the nest to do a little careful gardening. All through the erection of the tower the adult birds continued to carry prey to the nest. Even though we knew from past experience that they were easy birds to work, it was heartening to see both birds apparently unperturbed by the activity around them. Gardening finished, we moved the tower back two metres from the nest which just allowed us to bring the nest into focus using a 250mm lens with fully extended bellows. We did not lower the tower to move it back from the nest thus saving time and physical energy, but creating much nervous tension in the process.

The adult birds perched in a nearby tree while the tower was climbed and the hide entered. It was quite a relief to enter the hide as it blocked out the view of the forest floor so far below. The mental relief could be likened to that of the ostrich when it buries its head in the sand to avoid danger. Perhaps some of our critics were right when they said we were bird-brained. However, remaining anxieties soon dissipated as we concentrated on getting set up for the job in hand.

The female was not long in coming in with a phasmid. She alighted on a dry limb that extended for about two metres from and slightly below the nest. She was out of frame and focus at that point, and walked toward the nest till in focus where a shot was taken. As our main aim was to get landing shots while the under wing pattern is displayed, the camera was focussed on the end of the limb, but the next landing was close to the nest. It was obvious that the limb would have to be shortened and so restrict the area in which the birds could alight.

The male was next to bring in prey though he left immediately a chick grabbed the phasmid from him. He perched in a nearby tree and preened. With his back toward the hide he made an unforgettable picture as he stretched both wings out in the form of a magnificent fan. Unfortunately he was a little too far away for the lens in use and partly obscured by foliage.

The adults shared the feeding of the chicks fairly consistently. For much of the first day they came in three or four times per hour, but occasionally there was lull in the tempo during the afternoon when an hour could elapse between visits. The prey was phasmids and tree frogs. The latter were too large for the chicks to swallow whole, but each would try. Sometimes they would get the back legs down, but the bodies would prove too large, so they would disgorge them and have the adults tear them into manageable pieces.

The female brought sprigs of green leaves occasionally to lay in the nest. As the afternoon

The Pacific Baza in flight.

advanced it became very hot and sultry and she would stay on the nest for a while trying to shade them, but was only partially successful as they were too large to cover adequately. The heat caused them to move about a lot in an endeavour to find shade, and one marvelled how they managed to stay in the nest. On many occasions one or the other would appear to have passed the point of no return, so far did they get down the side of the nest, where a twig to which they clung only had to break or pull out of the flimsy platform of a nest to send them on their way to eternity, or wherever suicidally inclined chicks go.

It was quite intriguing to watch the female inspecting her brood after feeding them; she would peer to one side then the other of a chick's bill and finding any small piece sticking thereto, would remove it and feed it to that chick. Satisfied with that one she would repeat the inspection and performance with the other. An inspection of the nest for any crumbs ensued.

Due to the high humidity we sometimes had a problem with moisture accumulating in the flash heads and causing them to malfunction, but it was overcome by using the pressure pack we carried to remove moisture from the ignition system of our vehicle.

By mid-afternoon on the second day at the nest, conditions in the hide were most oppressive. A heavy bank of cloud was approaching from the north-east and the ever closer rolls of thunder helped convince us that we'd done a fair day's work. The storm struck shortly after we gained the shelter of 'Rockwood' homestead and it was difficult to believe that the chicks would not be washed or blown from that flimsy platform. The tower was somewhat protected by the pine forest in which it stood, but it was still a great relief to see it still standing next morning and the two chicks still on their nest.

A portrait of a Baza with prey - the body of a large insect.

In an effort to conserve battery power for the flash heads they were not switched on till the Noisy Miners and Friarbirds heralded the approach of the Baza. Occasionally a Miner followed a Baza to a nearby tree before returning to its territory which was apparently along the flight path of the Baza to its nest. Why other birds should be alarmed by the Bazas is difficult to work out as we've never known them to prey on fledglings. Perhaps it is just the general raptorial appearance

The feeding of the chicks is shared by both adults. The female Baza turns to leave the nest after having brought an insect.

The male, with his brighter plumage and deeper eye colour, pauses at the nest after bringing in a frog.

that raises anxieties; with a notorious nest robber - the Square-tailed Kite - resident nearby, a certain degree of paranoia is understandable.

By 19 December, we'd covered all the activities at the nest except those occurring for the first two hours of daylight, so the hide was entered at dawn. The female remained on the nest with a chick snuggled under each wing. We'd decided to remove most of the dead limb that morning and force the adults to alight within the framed and focussed area. She remained brooding till the saw was advanced to remove the limb. The saw,

The male Baza alights beside the nest with an insect.

The female Baza with leaves for nest lining.

The two fairly well fledged chicks.

with an extended handle improvised from a metre length of pine sapling, and badly in need of setting and sharpening, gave the chicks a rough ride and they protested loudly. Thereafter some lovely shots of both adults were obtained.

On our final day at the nest we discovered why the chicks possibly survived as often as they did in gales and storms. The wind velocity had been rising all morning and from time to time some fierce squalls buffeted the hide. The guy-ropes were tightened to steady it, but nothing could be done to stop the tree swaying. The male began

calling, presumably for the female, but when she did not appear he flew into the nest to protect the chicks, remaining till the squall, which lasted a few minutes, was over. The photographer also decided it was time to leave, as apart from apprehension for his own safety, there was little chance of getting good photographs with so much vibration travelling through the tower to the hide. Another squall hit as the hide was being vacated and the male again left his preening perch to go to the chick's assistance.

In 1979 we received information that another pair was nesting near Kogan, about fifty kilometres north of Dalby, Queensland. With Cec Cameron's help, the tower was set up with the hide just below nest level by 14 November. As the birds were still on eggs we played it very cautiously, taking two more days to raise it above nest level. The extra caution was probably unnecessary, but we couldn't afford to make any mistakes.

The hide was entered on the 16th to find the two chicks had just hatched, the one just emerging from the shell. The female had remained on the nest till the hide door was opened and then attacked vigorously, swooping to within a few centimetres of the intruder as he entered the hide. She was the first of the species to do so, but returned to the nest immediately the hide door was closed. After removing the egg shell she resumed brooding and remained so while the flash heads and cameras were placed in position only 1.6 metres from her.

Within an hour the male alighted with a large phasmid, the wings and legs of which had already been removed, and the female left the nest. The chicks with eyes still closed and bills agape

A typical clutch of Baza eggs.

awaited their first meal. Their heads appeared too large for their bodies and one could only marvel at where they got the strength to lift them, let along hold them in that gaping position for long periods. The male eyed the hide for a few moments before returning his gaze to the chicks now swaying with the effort of holding up their heads. He extracted a tiny amount of jelly-like substance from the body of the phasmid and placed his bill within that of the chicks. All the action was gentle, but positive,

The male Baza brings a frog to the nest where his mate broods their tiny chicks.

the chicks getting a little food each time as it was placed in the gullet or close to it. It was the only raptor chicks we've seen fed in this manner, all other chicks taking the food from the tip of the adult's bill when proffered. We acknowledge that we haven't seen many raptor chicks getting their first feed, so it is possible that some others are fed initially in that manner. It is, of course, common for many chicks other than raptors to be fed in that way. It was a very small meal, being over in less than two minutes, and they were not fed again during the occupancy of the hide which lasted another four hours. After two hours brooding the male was replaced by the female.

We had come up specially to observe and record on film the adults' roles during egg incubation. They were observed, but not filmed, of course, due to being too late setting up. The adults shared the incubation much the same as they did with the brooding of the chicks, and occasionally they both sat together on the nest. It was noted that the bulk of the feeding was done from soon after sunrise till noon. By the fourth day they were being fed tree frogs as well as phasmids. The male, by then, was doing most of the hunting and passing the prey to the female to feed the chicks. The days were hot and humid and it rained most evenings.

The rain had started with our arrival in the area as it had done on most occasions in past visits. Our hosts had been hand-feeding their stock for some months prior to our arrival on this occasion, and it was suggested we should come more often.

The male Baza looks on as the female prepares to feed the tiny chicks.

The nesting area of the Bazas near Kogan.

183

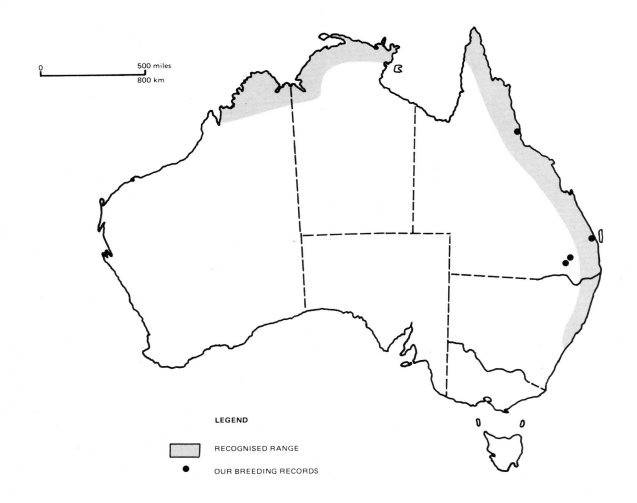

LEGEND

RECOGNISED RANGE

● OUR BREEDING RECORDS

0   500 miles / 800 km

PACIFIC BAZA   *Aviceda subcristata*

avis - bird (L); *cristatus* - crested (L).

OTHER NAMES: Crested hawk; Crested baza; Pacific or Crested lizard hawk; Pacific or Gurney's cuckoo-falcon.

LENGTH: 350 - 430mm. Female larger than the male.

WINGSPAN: Approximately 900mm.

DISTRIBUTION: Frequents forests and woodlands in coastal north-west, north-east and eastern Australia, south to southern New South Wales. Moderately common in the north, rare in the south. Mainly sedentary, possibly occasionally nomadic. Two Australian subspecies: *A. s. njikena* - north-west Australia. *A. s. subcristata* - eastern Australia. Also the Solomons, New Guinea and the Moluccas.

VOICE: Mellow double whistle **'wee-choo, wee-choo'**.

PREY: Mainly insects and insect larvae, also small lizards, frogs and possibly small mammals. Birds that we studied preyed on phasmidae and frogs, with occasional grubs and caterpillars.

NEST: A rather flimsy, shallow structure of light sticks and leaves, lined with green twigs and leaves. A typical nest is only 300mm in diameter and 150mm deep. It is usually on a horizontal branch that has some lateral growth to support the nest, and is 15 - 30 metres above ground.

EGGS: Usually two or three, though five has been recorded. They are rounded ovals, 43 x 34mm, fine and slightly glossy, pale bluish-white, occasionally sparingly marked with brown. We saw only one clutch - two eggs. Of three other nests observed two had two chicks the other three. The eggs are laid from September to March though usually October - December, which is the period indicated by our records. The eggs hatch in about 33 days and chicks fledge in 32 - 35 days.

184

OSPREY
Pandion haliaetus

This species is found throughout most of the world and varies only slightly in size and plumage colour. In Europe it is migratory, but in Australia sedentary, appearing to occupy its territory the year round and for the duration of its life. It is always close to water, generally along the coast and occasionally tidal rivers. We have located several nests along the coast of South Australia and on nearby islands and many along Queensland's coastal regions. One nest on a lagoon near Rockhampton was probably forty kilometres from the sea. It will nest in a variety of situations; on the ground, on cliff ledges, power-line pylons, rock pinnacles, tall dead trees, short ones and even on the roots of a fallen one. The same nest is often used for years, even decades, if not unduly interfered with during breeding. As with the Sea Eagle, new material is added each season and they become huge structures. All manner of flotsam may be incorporated in them.

Our favourite nest is in an old dead eucalypt tree within sight of the huge coal terminal at Hay Point, a few kilometres south of Mackay, Queensland. The nest is at a height of twelve metres, but the beautiful backdrop of mangroves and Louisa Creek give an impression of a far greater height.

As breeding starts in late May or early June it is the first nest we work each year. On the coast of South Australia breeding starts in September or October.

We first found the Hay Point nest in mid May 1974. A local resident, in reply to our enquiries, had told us he suspected a pair of Ospreys were nesting nearby. He had often seen them carrying sticks or fish toward a patch of timber bordering Louisa Creek. On one occasion he had been startled by a sudden clatter on his roof as a large stick had slipped from the talons of an Osprey passing overhead.

Within a few minutes of beginning our search we had found the nest - only about two hundred metres from our informers back door. We erected our tower but unfortunately it was, at that time, not high enough for us to see into the nest. However it was obvious the birds were incubating eggs, so we took a few photographs and left.

It was late in July before we were able to return. As we approached the nest, two juveniles flew, somewhat clumsily, to nearby trees. We erected the tower, but although the chicks soon returned to the nest and solicited for food, the adult birds did not come in to feed them.

185

An Osprey nest on a cliff face in South Australia.

An Osprey nest on an electricity pylon, North Qld.

Our favourite Osprey nest near Hay Point.

Two years later, in early July, we made another visit to the nest, to again find two chicks, about half fledged. We worked them only briefly, but were able to do some further work late in their breeding season as we were working on Brahminy Kites nearby. The two young Ospreys were flying and there was little to distinguish them from their parents. As they had not yet learned to fish, they were still being fed on the nest, and being relatively quiet birds to work we were able to get some very interesting film.

All fish are freshly caught and more often than not are alive and kicking when brought to the nest. Sometimes one is released before the young have grasped it and much confusion can occur as three or four large birds converge on a bouncing fish. On one occasion a fish flipped out of the nest, but no attempt was made to retrieve it. Weather conditions appear to have a marked effect on the efficiency of the Osprey's fishing. On calm days fish were brought in regularly and in greater numbers than on windy ones. Counts ranged from a low of two to a high of seventeen, but the total weight would not have varied as much as the numbers might indicate as the high tally was much smaller fish.

In mid July 1977 we found two healthy, well fledged chicks in the nest and a third chick dead on the ground below. This was the first time we had found evidence of a third chick although we are told that three eggs are often laid. It is possible this pair normally laid only two eggs, but apparently it is not uncommon for one egg of a clutch to fail to hatch.

On 12 July 1978 we set up again on this nest, where for the first time in five seasons it had only one chick. It was probably a little over a month old, being feathered on wings, back, tail and head, but breast and belly were still a dense coat of dark brown down. One apparently addled egg still lay in the cup of the nest. The chick wasn't being brooded during the day, but the female spent most of her time on the nest during the day and, presumably, all night.

186

The mangroves of Louisa Creek form a beautiful backdrop for the Ospreys' nest.

The hide was entered two days later, just before sunrise. The female, having left the nest as the tower was approached, kept voicing alarm calls as she circled above. About an hour later the female alighted with the first fish of the day. It was a pike about forty-five to fifty centimetres long. No landing shots had been taken in an endeavour to have her settled down quickly. Even so she was extremely wary and stared intently at the hide with facial and crown feathers ruffed. Occasionally she weaved her head from side to side still staring intently at the camera lens, while the chick solicited loudly to be fed. Eventually she relaxed a little and heeded the chick. She started on the still squirming fish by tearing off the upper beak-like snout which she passed to the chick. While the latter was manoeuvring it into a position suitable to be swallowed she tore off and ate the lower mandible herself. The head was torn apart next and fed to the chick, and so the fish was progressively eaten, nothing at all being discarded. When it was finished she peered around the nest to make sure nothing had been over-looked. Finding nothing she immediately started calling on the male for the next course. He wasn't long in responding with another pike only slightly smaller than the first. It was disposed of in exactly the same manner, starting on the snout and working toward the tail.

The windy conditions next day, coinciding with the tide peaking during the morning apparently made fishing unrewarding for the male. The hide, as usual, had been entered early and the female alighted on the nest around 08:00 without a fish and was immediately solicited by the chick. It soon realized she was empty-handed and returned to its preening. The female called almost incessantly for the male and from time to time walked around the nest edge scanning the sky for a sign of him. The chick preened almost continuously except for short breaks to back toward the edge of the nest, at roughly half hour intervals, to eject a stream of excreta well clear of the nest.

The Ospreys with their well fledged chicks.

At 10:30 both the female and her chick were calling, and from their obvious excitement it was evident that the male was on his way in, so the flash was activated in readiness for a landing shot. A shot was taken as he appeared within the framed area, and when the mirror was cocked again he was perched lightly on her back, but was airborne again before the film could be advanced for the next shot. He had brought nothing in, though he'd been searching the estuary since daylight. The high tide and choppy sea surface had probably prevented him from seeing anything. He'd no doubt heard her every call as it was only a few hundred metres to the estuary where we had our camp, and we could always hear her and the chick calling. Possibly he'd come to the nest to let her know he was still hunting and not philandering or perched preening himself. She was mollified only for a short time and her obviously empty crop and that of the chick's, which now showed a cleavage line from throat to belly, gave her no peace of mind. She started calling again and was joined by the chick. Although it's crop was long since empty its stomach and bowel obviously wasn't as it had backed to the edge of the nest and appeared to be aiming at the camera. It fell short, but it was just as well we weren't working as close as we sometimes do with smaller species.

At 13:00 the male brought in the first fish of the day and a good landing shot was obtained, but he didn't stay long enough for a second one to be taken. Whether it was his mate he was afraid of or the hide we weren't sure, but would guess it was the hide. The fish was a substantial one, species unknown. It was also a very tough skinned one, going by the effort needed by the female to tear a hole in it. With much straining and head twisting she eventually managed to tear a small opening through which she extracted small bits of fish to feed a very impatient chick. After one and a half hours toiling, all that remained was a long strip of very tough skin still attached to the bony head held under her talons. For the next half hour she gulped down that length of leathery skin till her bill reached the bony head. At that point she

A typical clutch of Osprey eggs.

188

The Osprey in flight.

The Osprey alights on a vantage point above its nest.

would clamp her bill and strain upwards, but the result would be the slow extraction of the skin from her crop via her clamped bill. The few remaining scales on the skin were slowly removed and swallowed with each extraction. She had at least as much application toward her task as Robert The Bruce's spider and after a score or more attempts managed to tear the skin from the head. No sooner had she finished gulping it down than she was calling on her mate again.

As the photographer had, hopefully, a couple of good landing shots and one of the chick with a very full crop, he decided to call it a day after eight hours in the hide. He climbed painfully, but carefully, down to *terra firma*.

On several occasions we'd watched the male's fishing techniques. Sometimes he would perch on top of a cliff over-looking the sea from where he would make a steep dive into it to take a fish. On rare occasions we'd seen him taking them in a manner similar to that of the Sea-Eagle, swooping down to take the fish with striking talons, his body clear of the water. His third and most usual method was filmed on the day following the eight-hour session at the nest. It was at his favourite spot where Louisa Creek meets the Pacific Ocean,

and as it was low tide it was his most rewarding period. It was a cold day with an easterly wind and he soared slowly into it at perhaps a hundred metres. At times he appeared stationary and may have spotted something, but then moved on to hover or hang on the wind again. Suddenly he dropped to half his height to hang stationary for a second or two before plunging feet first into the water. He disappeared from our view beneath the surface, but a few seconds later we spotted him wallowing on a small wave. Moments later he had somehow flapped himself clear of the water and flew toward us and the cliffs. He had a relatively large fish in his talons and appeared to be making heavy going; suddenly he shook himself to shed some of the water from his feathers and in doing so lost much of the height he'd gained. Without the weight of the water he quickly regained height and passed over our heads to be swept swiftly upward without apparent effort by the wind currents near the cliff face. From his position above the cliffs it was all downhill to the nest. From where we stood we could hear the female and chick excitedly calling him home. It was also time for us to head for home.

Osprey approaching nest with fish

The scene from our camp at the mouth of Louisa Creek, looking toward the Hay Point coal loading terminal.

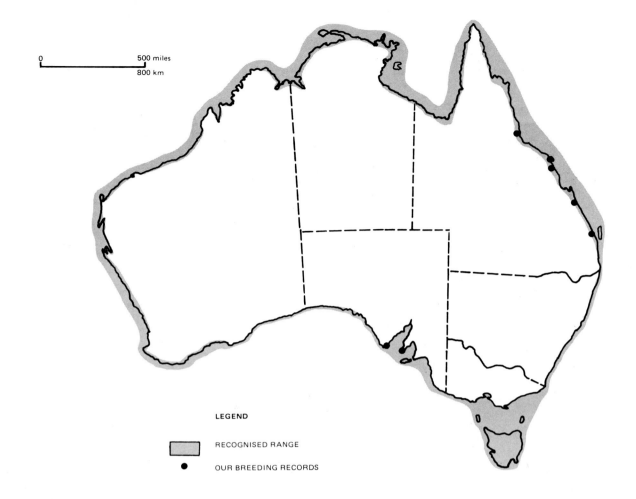

0 ——— 500 miles
800 km

LEGEND

▭ RECOGNISED RANGE

● OUR BREEDING RECORDS

OSPREY *Pandion haliaetus*

Pandion - mythical King of Athens who was turned into an Osprey; **hals** - sea (Gk); *aetos* - eagle (Gk).

OTHER NAMES: White-headed osprey; Fish-hawk.

LENGTH: 500 - 630mm. Female slightly larger.

DISTRIBUTION: Around entire Australian coast near shores, islands, estuaries; sometimes well inland along larger rivers. Moderately common in the north, rare in the south; sedentary. The Australian subspecies *P. h. cristatus* extends to New Guinea, Indonesia and South East Asia. Other subspecies range over most of the world.

VOICE: Usual call is a quavering **'pee-ee, pee-ee'**. When disturbed from the nest, or at the invasion of another raptor the call is a short, sharp **'tchip-tchip-tchip'**. The food call is a high-pitched **'pseek-pseek'**.

PREY: Although in the Northern Hemisphere ospreys occasionally take birds, frogs and crustacea in Australia they appear to feed exclusively on fish.

NEST: A very large structure of sticks and drift-wood, lined with seaweed, placed in a tree up to 30 metres above ground; on a cliff-face or rocky pinnacle, or (particularly on off-shore islands) on the ground or rocks. Man-made structures are also used. We saw several nests on transmission towers in Northern Queensland.

EGGS: Usually two or three, rarely four, 61 x 44mm. They are rounded ovals, coarse and gloss-less, white or cream white, boldly spotted and blotched with chocolate and red-brown, with underlying purplish-grey markings. Of 7 clutches we recorded: 4 consisted of three eggs and 3 were of two. The eggs are laid from May to September in the north, and September to November in the south. Our records show egg-laying in late May - early June in north Queensland, and early October in South Australia. Incubation time is 33 - 35 days and chicks fledge in about 50 days.

RED GOSHAWK
Erythrotriorchis radiatus

The Red Goshawk is, without doubt, Australia's rarest and least-known diurnal raptor. Very few people have seen it and breeding records are extremely rare and mostly sixty to one hundred years old. It frequents open forests and wooded areas of northern and eastern Australia, apparently favouring the more remote and rougher regions, and so probably contributing to the paucity of records. It appears that the Red Goshawk, although always very rare, formerly had a wider distribution, extending as far south as Sydney.

Although this species was sometimes called the Red, or Rufous-bellied Buzzard, it was first described as a Falcon, but later placed in a genus of its own - *Erythrotriorchis*, *erythros* meaning 'red'; *triorchis* meaning 'bird'. *Radiatus* means 'barred'.

In those early days, egg and skin collection was the main preoccupation of most who sought it, and as a result, data generally are few and meagre. Quoting from the **Emu**, H. G. Barnard observed:

> *Erythrotriorchis radiatus*. Only a few of these fine birds are seen. A nest found early in September contained one hard set egg. A second nest, with two fresh eggs, was found a few days later, and under this nest lay the remains of a Nankeen Night-heron *Nycticorax caledonicus*. On visiting the nest exactly four weeks later, two fine eggs, evidently laid by the same pair of birds, were taken. While taking the eggs and male bird brought a Naked-eyed Partridge-Pigeon *Geophaps smithi* with which to feed his mate.

Quoting from the same source, 24/5/13: 'Crop of *Erythrotriorchis radiatus* contained remains of large grasshoppers'. [13]

Prior to 1977, when we first began seriously looking for this species, we had just one breeding record which we considered recent enough to be worth pursuing. In November 1974 Norm Favaloro of Mildura contacted us with news of a nest of this species he had found on Cape York Peninsula. The chick was fully fledged and out of the nest at the time, so we thought it not worthwhile travelling the eight thousand kilometre round-trip involved, until some future date. Unfortunately Norm's recollection of how far it was from a river or track junction was vague. 'Anything from two to five miles', although helpful left a lot of country to be searched.

Our first sortie was made in 1977. Our Falcon utility didn't handle the corrugated roads at all well. Any attempt to exceed forty kilometres per hour would have been impossible over much of the track north of Mareeba. It was definitely four-wheel-drive country and we had nothing to spare, traction-wise, when climbing some river and creek banks. It was an experience, but the tension created by the limitations of our vehicle, which as always was over-loaded, didn't make it an enjoyable one. We were intrigued by the number and size of the termite mounds. There were literally millions of them. Some stood four metres high and must have weighed a tonne. In numerous places the magnetic type, stood like head-stones in a vast cemetery. Our search that year convinced us it wasn't going to be an easy bird to find. Driving home with very little time out for sleeping or eating we covered the four thousand kilometres in seventy-three hours elapsed time.

It was while embarking on this trip that Henry Nix of the C.S.I.R.O. Division of Wildlife Research told us of a nest he had found in the Shoalwater

On Cape York Peninsula the termite mounds stand like headstones in a vast cemetery.

A mound of the magnetic termite. The narrow ends always point North and South.

The cathedral type mounds are sometimes massive structures.

Bay Military Training Area in Queensland in October 1971. On our return we wasted no time in obtaining access to the area, but the vast expanse of woodland and forest would have taken many weeks to search thoroughly, so after combing what we thought was the most likely area, we deferred further efforts till the following year.

While studying the little information available, it came to our notice that there had been, over the years, several sightings of the species along some of the rivers of the Northern Territory; in particular along the McArthur and South Alligator Rivers, so late in July 1978, we headed north.

August found us camped on the banks of the McArthur River, just upstream from the small town of Borroloola. Here the river was under tidal influence, rising and falling over a metre each day, and we occasionally saw small sharks drifting past our camp. Of more concern, however, was the five metre long salt water crocodile reported to be resident nearby, but after walking all day in the oppressive heat, it took more than a crocodile to keep us out of those inviting waters. However, after two weeks, with no Red Goshawk but severe gastric trouble, we moved on to conduct a brief but unsuccessful search along the Roper River, and from there to the upper reaches of the South Alligator. Here once again an extensive search was unsuccessful so we headed south to Alice Springs and then east back into Queensland.

During our sojourn through the Northern Territory a Brisbane-based newspaper carried a story of our project and a request for information on the Red Goshawk. This resulted in John Young contacting us with news of a pair of Goshawks he had seen on several occasions. We received this information while in Maryborough. Nineteen hours and fourteen hundred kilometres later we were talking with John. Next morning we began our search and within an hour we had our first definite sighting of this elusive species. Things looked very promising. However, after a week of constant searching we had no further sightings and it was time for us to return home.

Fortunately, as well as the memory of that one, all too brief, glimpse of the Red Goshawk, we had, thanks to John, a much wider knowledge of the local birds, and more importantly, further information about our quarry. He told of finding in the early 1960's, the Red Goshawk nesting in the Narren River - Narren Lake area of northern New South Wales. He had records of several nests including one containing three eggs, the normal clutch being one or two. Although he had made no extensive study of the birds, he believed them to be feeding largely on water fowl. He was also able to give us information on nest site and construction.

With only this species to find and film by 1980 we were able to plan a prolonged campaign to get it. We set out with the intention of searching every area in which the species had been known to have nested in the past half century. They included the Narren Lake and river, Shoalwater Bay, the Cape York Peninsula and Murphy's Creek near Toowoomba.

We had made a short but unsuccessful foray into the Narren River-Lake area in 1978. In April 1980 we again visited the area and found the lake dry and the river just a series of shallow mud holes teeming with carp. The whole area had been declared a drought area for a considerable time and carried little or no stock. Even the kangaroos and feral pigs were too weak to hop or run. It was definitely no longer Red Goshawk country.

We next searched an area around Rockhampton where a pair were reportedly seen a few weeks earlier. The lagoons and surrounding area appeared suitable habitat but we made no sighting. Some time was spent searching along the Condamine River and nearby creeks and lagoons near Chinchilla. A single bird had been reported for some years but we didn't sight it.

We then contacted a former resident of Murphy's Creek, Max Miles. He had known of three nests in the area forty-two years earlier. He spent a few days overlooking a valley there in 1978 and had sighted a pair flying close to where they had once nested. Max travelled from his home north of Brisbane to take David and ourselves on a conducted tour of the old nest sites that he'd known. We thought it perhaps our strongest lead and we searched a wide area. Although we got no sighting of the species we did, however, get much fitter physically. Fires and logging in recent years may have driven the Goshawks further back into the ranges, but there is a fair chance that they could still be in the area we searched, as it was rugged and quite heavily forested. It would have been very easy to miss the quiet type of bird it turned out to be.

Cape York Peninsula was next on our itinerary as we couldn't get permission to enter the military training area of Shoalwater Bay till 8 September, when the military would grant us ten days and an escort. We left home and headed north again to Toowoomba to spend a few more days in the Murphy's Creek area before moving on to spend a couple of days searching north of Yeppoon where Dr. Ray and Norma Channels had seen a pair in recent weeks. There was plenty of suitable habitat, virgin forest, woodlands, lagoons and creeks teeming with bird life, but no Red Goshawk.

roughest track we'd encountered in six years around Australia. From Mt. Carbine to the Wenlock River was a hard day's grind. Some rivers and creeks had a small flow but presented no problems. The previous Wet Season had been a rather dry one and most rivers were lower than normal.

We started our search at Wenlock, which is the site of an old gold mine. There is no town but going by the amount of old mining plant it must have employed quite a team at some time. We were told that an argument between the mining partners ended in one being shot. There are two graves there, one with a rough-hewn wooden cross bears the inscription: **Wm L. Stanley Died 1957 aged 57 yrs;** the other's head-stone reads: **In Memory of Thos. Power died 10/2/30.**

We had to use compasses to keep on a steady course while searching. We were satisfied that any country searched was done thoroughly. Using

One of the better creek crossings on the track up the Cape.

A couple of days were spent at Louisa Creek where we searched for and found our Brahminy Kite nesting. We set up the tower at the nest so that the birds could get used to it in our absence up the Cape. We didn't consider our chances of finding the Goshawk on the Peninsula good enough to take the tower and risk damaging it in the rough terrain north of Mareeba.

We came to the end of the bitumen just north of Mt. Carbine. Our diesel powered Toyota light truck handled the conditions much better than our Falcon had three years earlier, but it was still the

a track as a baseline we would walk out for an hour and then back on the opposite leg, roughly one hundred paces from the outward leg. By the end of the first day we'd covered perhaps a thousand hectares and had perhaps half a million left to do and only a week before we would have to leave for Shoalwater Bay to keep our appointment with the military. It appeared a hopeless task but we knew there was little chance of success sitting at home or in our camp.

On 1 September, we'd breakfasted and were on the move by 06:30. We had completed the first

The female Goshawk stands at the back of her nest.

outward leg and were almost back to the base-line when a large stick nest was spotted on a large horizontal branch of a Bloodwood and about fifteen metres above ground. As we examined it through binoculars a bird stood up and peered down for a few moments before flying leisurely to a nearby tree. With its back to the breeze it appeared to have a crest which, coupled with its barred tail gave the impression of a large Baza. With the arrival of the male bird there was no doubt that we'd found the Red Goshawk. His rufous breast and thighs gleamed salmon pink in the morning sun, an optical illusion no doubt, as he didn't appear as luminous with his breast in shadow. We went back to our camp, walking on air.

By 12:00 we were on our way to get our tower 1500 kilometres to the south. It meant four days of steady driving. The fourth was a marathon of 14 hours, much of it in second and third gear as we negotiated the rough tracks, at times on top of the range, at others descending or ascending the mountain trails. The steady speed and extra weight with the tower atop emphasized the impact

of the corrugations It seemed it would rattle our teeth loose. We shook like jellies and the dust caked thick over everything. There's a saying that anything one gets for nothing is worth nothing, so on that basis we think this bird should be worth a fortune. It was dark when we reached our camp on the Wenlock River. We needed no rocking to sleep that night.

Next morning we drove to within a few hundred metres of the nest site and walked in. The bird was not on the nest. We walked around for a few minutes searching the tree-tops while our apprehension increased with every passing second. It was just unbelievable. Convinced that she must have deserted we felt flat and suddenly very weary. We had just decided we may as well examine the nest and its contents, if any, when we heard the female coming. We breathed freely again as she alighted in the nest-tree. We walked about two hundred metres away where we could keep the nest under observation. She was apparently keeping us under observation too, as she flew out to perch in the tree under which we sat, and peered down at us. Satisfied that we meant

The single Red Goshawk egg.

her no harm, she soon returned to the nest and resumed incubating.

We set the tower up against the tree which had been impossible to climb owing to its large girth. The lower part of the nest was constructed of substantial sticks up to fifteen millimetres in diameter, while the upper section was a deep saucer of finer twigs, well-lined with green leaves. We believe the base of the nest was probably that of a Black-breasted Buzzard, a pair of which we saw several times nearby.

The single egg was photographed in the nest and checked for state of incubation by placing it in a tin of tepid water hauled up on a light line. We'd come prepared with water in a thermos flask. The egg was well set and hatching estimated (or 'guesstimated'), to occur near the middle of September. The tower was placed about eight metres from the nest-tree and left at its basic height of seven metres. All this took just thirty minutes and

The Red Goshawk in flight.

we observed the female return to the nest five minutes later. Later in the day we moved the tower to 4.3 metres from the nest and raised it to thirteen metres. The female accepted it without hesitation so it was raised to just above nest level next morning.

The hide was entered early on Sunday 7 September and a few shots taken as the female sat incubating. She only left the nest for a few moments as the hide was entered. It appeared she would be an easy bird to work. On Monday she didn't bother to leave the nest as the tower was climbed and the hide entered only four metres from her. She looked a little like a Square-tailed Kite as she sat incubating, and certainly had the same reaction to human intrusion as one. By that evening we decided we may as well return home as there was little to be gained photographically till the chick hatched.

The male's role entailed the hunting of prey (other birds) which he brought to a nearby tree. He would call her off if she hadn't already seen him coming and she would fly out to snatch the prey from his talons. He would follow her to whatever tree she fed in, and quite often alighted on her back for a moment. She sometimes mantled over the prey and kept up a noisy protest as if she was expecting him to try and take it from her. His action appeared as playful teasing, but was most probably a sort of mock or symbolic copulation - pair bonding behaviour.

We left for Cairns in the evening, getting about two hundred kilometres on our way that night, and arrived in Cairns next afternoon. We had decided to fly home, but by the time we'd got our plane tickets we had just over an hour to park our vehicle and get cleaned up. Luckily we found an old friend, Sandra Clague, home. We parked our vehicle in the Clague garden and showered, (without clogging the drains) while Sandra prepared a 'cuppa' before driving us to the airport. At 17:05 we were flying south and could relax.

The tower at the Red Goshawks' nest.

In Melbourne we took the opportunity to get our films processed. We were told they would be ready in four hours and we duly returned with the sweet anticipation of at last seeing a reward for our labour. All those thousands of kilometres of driving, and hundreds walking, under all manner of conditions would be rewarded by those photographs, the first ever taken of the species at the nest. We were in for a severe shock. The first envelope contained strips of blank film, the second likewise, the third one shot of the bird, and the fourth two shots of the tower and habitat taken with a 90mm lens. The 360mm lens shutter had failed to operate after the first shot was taken. We'd both used the same lens and camera instead of changing them for our individual ones. We flew home and in spite of our deep disappointment with the films it was nice to be home after twenty-seven hectic days of travelling and searching.

In less than twenty-four hours we were on our way back. We had to get some further photographs and film as there was always a chance of the egg hatching and the chick dying before we got back. David flew out with us, and of all the passengers on that plane, only his and our luggage was missing when we got to Cairns. It seemed the gremlins were still at work. It might be on the next plane we were told. We decided we may as well get fuelled and victualled in the meantime. We found all three batteries on our vehicle flat; the refrigerator had been left switched on when we'd left Cairns in haste a few days earlier. What would be next? It proved that adage that troubles seldom come singly. We eventually started the

vehicle, fuelled, victualled and awaited pessimistically for our luggage. It arrived and we withdrew the writs we'd taken out mentally against the airline. By 13:40 next day we had David in the hide on the Red Goshawk. We planned to share the hide in rotation for the next week.

During the few days we had the nest under observation in the incubation period, there was very little action. The female would leave the nest once or twice a day to bring a sprig of eucalypt leaves to spread on the nest. Her absence from the nest to feed was between twenty and thirty minutes once a day. After snatching the prey from the talons of the perched male as she swept past him, she would keep up her noisy calls for some time. The call was somewhat like those of the *accipiter* genus, though generally not as rapid or shrill, except when alarmed by the close approach of another raptor or corvid, when the alarm was a very shrill and rapid staccato chatter.

In the afternoon of 15 September the female would occasionally give a start, and raise herself slightly to peer at the egg. Some cheeps seemed to be coming from the direction of the nest, but one couldn't be certain as there was much twittering from small birds around about. At one stage she turned the egg with her bill but no holes or cracks could be discerned by the occupant of the hide four metres away.

Anticipating that the egg was about to hatch, the cine camera was taken into the hide next morning. From the posture of the female it was obvious she was brooding the chick. She didn't wriggle down low in the cup of the nest as she had done when incubating. She eventually stood up to reveal the snow white chick. She took the two pieces of egg shell in her bill and flew off to dispose of them. The sound of the cine camera running didn't inhibit her in any way.

The chick wasn't fed that day but that didn't worry us, as many chicks don't get fed during the first day after hatching. However we were puzzled when forty-eight hours had passed without our seeing the chick being fed. We'd been sharing the hide from sunrise till sunset, and, although the chick appeared healthy and the female unperturbed, we were beginning to wonder if our

The female Goshawk returns to incubate her egg.

The male Goshawk perches in a nearby tree.

presence was upsetting the male. This seemed unlikely as, prior to the chick hatching, he seemed unworried by us. Never-the-less we decided to lower the tower and keep the nest under observation till dark.

Darkness comes quickly after sunset in those latitudes, and it was almost dark when the male arrived with prey. Apparently, since the chick had hatched, it had been fed in the twilight between sunset and dark and quite possibly in that period before sunrise. We later found there was no set period for the male to arrive; he might turn up in the middle of the day or before sunrise. Two feeds a day at this stage seemed about his normal routine. It had so happened that he'd arrived in the twilight during those first few vital days and in so doing had fooled and worried us, but we doubt if his behaviour had been altered by our presence. He appeared more curious than alarmed by us walking or standing below him while perched relatively low in nearby trees.

The crepuscular behaviour of the male brought to mind a possible sighting we had of this species in the Shoalwater Bay Military Training Area in 1977. We had finished our evening meal and were lying back idly watching the last rose-tints fading from the western sky, when a relatively large bird flew overhead and quickly disappeared in the gloom to the east of us. Its silhouette didn't

A Rainbow Lorikeet - popular in the Goshawks' diet.

accord with any raptor that we knew, but never having heard of this species being crepuscular we didn't seriously think of it being the one for which we were searching. As it was within a couple of kilometres of where Henry Nix had found a nest in 1971, and was heading in that direction, we now feel certain it was the male Red Goshawk on its way home with the evening meal.

The prey was mostly Rainbow Lorikeets, of which there were many in the area. The female removed many of the feathers before she brought them to the nest but there were still enough colours left to identify them. The male never came to the nest during the period we had it under observation up to 19 September.

The female Goshawk feeds the three day old chick from a Rainbow Lorikeet.

Many of the reports we'd received emphasized the alarm this species caused among other birds. We found this so, well away from the nest where he presumably hunted, but neither bird caused much alarm close to the nest. Sometimes a slight rise in the twittering of other birds alerted the female that the male was on his way in, but on other occasions the first intimation that she or the photographers had was a couple of calls from him as he approached. His hunting ground appeared to be at least three kilometres from the nest but we never actually saw him take prey. His presence in the area had the other birds alarmed so we assumed it was where he hunted.

On 19 September we left the area as David was due back at his surgery on the following Monday and there wasn't a lot we could do photographically. The only other bird we found nesting in the area was the Lemon-bellied Flycatcher, which was nesting close to the Goshawk. Its tiny nest must be the smallest of any Australian bird. We did not work on it as we were worried about the male Red Goshawk's behaviour at the time.

We returned to the nest area after a three week absence to find a fire had burned through the area and all guy ropes tied to steel pegs had been burnt. Luckily we had enough tied to trees and well clear of the ground that had escaped the fire and kept the tower upright. A six metre section of tower not needed for the Goshawk nest and left lying on the ground, had been stolen. The thieves must have envisioned other uses for it, as a fold-up hide with the canvas burnt from it was still there. The fire had no doubt made the section visible from the road. Its loss was a severe blow, as it inhibited our ability to work other species now starting to breed. We were loath to remove the tower from the Goshawk's nest to work other species as that could have had an unsettling effect on the Goshawks.

We were surprised to find the chick hadn't advanced as much as we had expected in our absence. It was only lightly feathered on the back and wings, the rest of it was still white down.

The male had probably been coming to the nest occasionally while we were away as he appeared on the nest without prey on the second day we were back. He appeared very nervous and flew off after a shot was taken. The next day he appeared on the nest again with a Lorikeet, but the cine noise frightened him. The starting of the cine was a reflex action as it was not realized it was the male till the female flew in moments later with a Laughing Kookaburra. It would appear that she

After a day's feeding, only the head and back-bone of a Kookaburra remain.

had taken it herself as it was the only time we had seen anything larger than a Lorikeet brought to the nest. His arrival at the nest with a Lorikeet only moments before also indicated that she had perhaps resumed hunting. After feeding the chick she flew off with the substantial remains. When she returned some time later without it we assumed she had eaten it. Our assumptions proved wrong, as she made two more visits to the nest later in the day to feed the chick from it till only the bill, head and backbone remained. Even the brains had been removed from the head.

Because of their massive legs, feet and talons we had expected the Goshawks to be taking much more prey of this size. They are reputed to take small mammals, reptiles and birds such as Night-herons and Ducks, but the only other sizeable bird that may have been taken was a Blue-winged Kookaburra, the remains of which we found beneath the nest-tree. We feel that had the chick been more advanced the female would have been hunting more and probably taking larger prey.

On 15 October the tower was moved in to two and a half metres from the nest for some closer photography. It had no effect on the female as she never bothered to move off the nest when we were entering or departing the hide. The next day was an extremely hot one and she shaded the chick with slightly extended wings early in the after-noon. The nest was always in shade till about noon or a little later and she generally perched a little higher up the nest-tree unless shading or feeding the chick. If we got to the nest before dawn we'd find her brooding the chick, but she always had fresh leaves on the nest before sunrise.

We found two and a half metres from the nest a little tight for the 250mm lens sometimes, especially while the chick was feeding. The chick would stand on one edge of the nest while the female would stand on the opposite edge, and both would have to stretch forward for the food to be taken. This often resulted in the chick over-balancing into the cup of the nest. They didn't always feed in this stand-off manner, but why they should do it as often as they did remains obscure.

We moved the tower out later to three metres from the nest in an endeavour to obtain landing shots of the female. Our final two days were spent

The female Goshawk uses her body and wings to shade her chick.

trying to get cine coverage of the prey change-over and the flight of the female to the nest, but it wasn't very successful for two reasons; the tree density and the inability to rotate the hide.

Our camp on the river was one of the best we'd had in more than six years of searching for and working birds of prey. We were able to vary our diet with an occasional feed of fish. The fish also attracted the fresh-water crocodiles, resident in the same pool. A 1.3 metre specimen was caught and carried up the bank to recuperate in the sun. Thinking that a little wetting down might be appreciated by the reptile, it was approached to within a couple of metres with the intention of throwing a billy of water over it. The crocodile misinterpreted the intentions apparently, as it leapt into the air and landed stiff-legged and open-jawed facing its would-be benefactor. The crocodile took advantage of the momentary state of shock it had inflicted and raced down the bank and into the waterhole.

The freshwater crocodile is relatively harmless, though they do have very sharp teeth. To our surprise we found that a lagoon a few hundred metres from our camp was home for 'Old Charlie', a four and a half metre saltwater crocodile. We had been under the impression (a mistaken one, apparently) that we were too far inland from the coast for that species. However, he didn't worry us, nor we him.

The chick was almost five weeks old when we left the area on 19 October, and we estimated that it would be close to eight weeks old when it left the nest. The accumulation of sticks on the ground

The Wenlock River during the Dry season.

The freshwater crocodile we caught in the waterhole.

Our camp at a waterhole on the Wenlock River.

below the nest led us to believe these birds had nested there for several years, and feeling confident that they would continue to do so, we left our tower at nearby Wolverton Station as we made our way south.

Many hours later, as we completed that last bone-jarring descent down off the range, and the thumps and rattles gave way to the sweet hum of tyres on bitumen, our thoughts turned to home and family still three and a half thousand kilo-metres to the south. But as we drove on through the long, warm night there were other voices calling: there was the gentle ripple of the river below our last camp and the rustle of leaves in the huge melaleuca above, the raucous screeching of the gaudy Lorikeets as they flashed across the cobalt sky, and the mournful call of the Frog-mouth as we lay beneath the brilliantly starlit sky in the cool of the night. We heard them all, and rejoiced in the knowledge that we would be back.

The female Goshawk feeds her five week old chick.

The female Red Goshawk shows her massive legs and lethal talons.

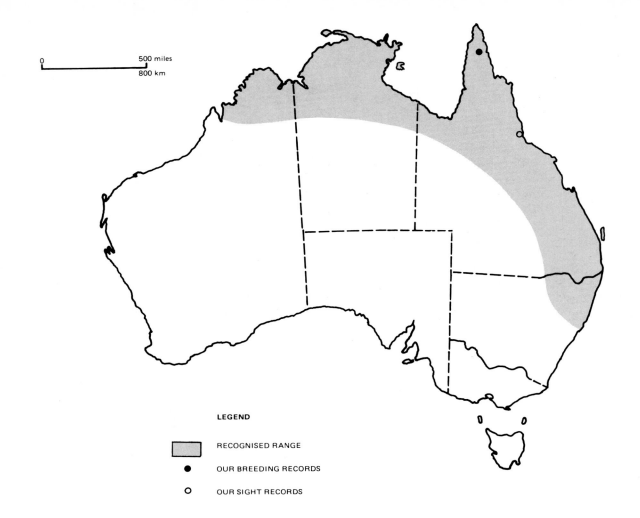

LEGEND

	RECOGNISED RANGE
●	OUR BREEDING RECORDS
○	OUR SIGHT RECORDS

RED GOSHAWK *Erythrotriorchis radiatus*

erythros -red (Gk); *triorchis* - bird (Gk); *radiatus* - barred (L).

OTHER NAMES: Red buzzard; Rufous-bellied buzzard.

LENGTH: 470 - 580mm. Female larger than male.

WINGSPAN: Approximately 1200mm.

DISTRIBUTION: Very rare, in open forest and woodlands in coastal and sub-interior regions of northern and eastern Australia, south to northern New South Wales. Probably sedentary. Not found outside Australia.

VOICE: Similar to the Australian goshawk, but slower and less shrill.

PREY: Chiefly birds, but also small mammals and reptiles. We have seen only birds brought to the nest as prey, Rainbow Lorikeets on most occasions, but also some unidentified passerines and on one occasion a Laughing Kookaburra.

NEST: A large, rough structure of sticks and twigs, lined with green leaves and placed high in a tree. The nest of another species may sometimes be added to. The nest we worked was about 750mm in diameter and 380mm deep, with a centre cup 200mm in diameter and 75mm deep. The lower part of the nest was constructed of substantial sticks up to 15mm in diameter, while the upper section was a deep saucer of finer twigs and leaves. It was placed in a horizontal branch in an open situation, 15 metres above ground. The base of the nest may well have been that of a Black-breasted Buzzard some years earlier.

EGGS: One or two, rarely three, 56 x 46mm. They are dull bluish-white, rounded ovals, unmarked or sparingly smeared or blotched with brown. We have found only one clutch - a single egg. The breeding season is thought to be from April to September in northern Australia, and August to November in the east. Our record indicated laying in mid-August. Fledging of chicks would appear to take from 50 to 55 days.

GLOSSARY OF SCIENTIFIC NAMES

Emu	*Dromaius novaehollandiae*	Australian Bustard	*Ardeotis australis*
Australian Pelican	*Pelecanus conspicillatus*	Inland Dotterel	*Peltohyas australis*
Rufous Night-heron	*Nycticorax caledonicus*	Silver Gull	*Larus novaehollandiae*
Grey Teal	*Anas gibberifrons*	Caspian Tern	*Hydroprogne caspia*
Osprey	*Pandion haliaetus*	Domestic Pigeon	*Columba livia*
Black-shouldered Kite	*Elanus notatus*	Crested Pigeon	*Ocyphaps lophotes*
Letter-winged Kite	*Elanus scriptus*	Partridge Pigeon	*Petrophassa smithii*
Pacific Baza	*Avideda subcristata*	Galah	*Cacatua roseicapilla*
Black Kite	*Milvus migrans*	Little Corella	*Cacatua sanquinea*
Square-tailed Kite	*Lophoictinia isura*	Rainbow Lorikeet	*Trichoglossus haematodus*
Black-breasted Buzzard	*Hamirostra melanosternon*	Crimson Rosella	*Platycerus elegans*
Brahminy Kite	*Haliastur indus*	Red-rumped Parrot	*Psephotus haematonotus*
Whistling Kite	*Haliastur sphenurus*	Barking Owl	*Ninox connivens*
Galapogos Hawk	*Buteo galapogoensis*	Frogmouth	*Podargus sp.*
Harris' Hawk	*Parabuteo unicinctus*	Laughing Kookaburra	*Dacelo gigas*
Brown Goshawk	*Accipiter fasciatus*	Blue-winged Kookaburra	*Dacelo leachii*
Collared Sparrowhawk	*Accipiter cirrhocephalus*	Richard's Pipit	*Anthus novaeseelandiae*
Grey Goshawk	*Accipiter novaehollandiae*	Red-capped Robin	*Petroica goodenovii*
Red Goshawk	*Erythrotriorchis radiatus*	Lemon-bellied Flycatcher	*Microeca flavigaster*
White-bellied Sea-Eagle	*Haliaeetus leucogaster*	Willie Wagtail	*Rhipidura leucophrys*
Wedge-tailed Eagle	*Aquila audax*	Chestnut-crowned Babbler	*Pomatostomus ruficeps*
Australian Little Eagle	*Hieraaetus morphnoides*	Thornbill	*Acanthiza sp.*
Spotted Harrier	*Circus assimilis*	Little Friarbird	*Philemon citreogularis*
Marsh Harrier	*Circus aeruginosus*	Moisy Miner	*Manorina melanocephala*
Black Falcon	*Falco subniger*	Yellow-throated Miner	*Manorina flavigula*
Peregrine Falcon	*Falco peregrinus*	Zebra Finch	*Poephila guttata*
Australian Hobby	*Falco longipennis*	Common Starling	*Sturnus vulgaris*
Grey Falcon	*Falco hypoleucos*	White-breasted Woodswallow	*Artamus leucorhynchus*
Brown Falcon	*Falco berigora*	Magpie Lark	*Grallina cyanoleuca*
Australian Kestrel	*Falco cenchroides*	White-winged Chough	*Corcorax melanorhamphos*
Gyr Falcon	*Falco rusticolus*	Pied Butcherbird	*Cracticus nigrogularis*
Quail	*Coturnix spp.*	Australian Raven	*Corvus coronoides*
Spotted Crake	*Porzana fluminea*	Little Raven	*Corvus mellori*
Eurasian Coot	*Fulica atra*	Little Crow	*Corvus bennetti*
Brolga	*Grus rubicundus*		

Crocodile, Freshwater	*Crocodylus johnstoni*	Lizard, Bearded	*Amphibolurus barbatus*
Crocodile, Saltwater	*Crocodylus porosus*	Mouse, House	*Mus musculus*
Dingo	*Canis antarcticus*	Rabbit	*Oryctolagus cuniculus*
Dunnart, Fat-tailed	*Sminthopsis crassicaudata*	Rat, Long-haired	*Rattus villosissimus*
Frog, Tree	*Hyla sp.*	Skink	*Egernia sp.*
Goanna	*Varanus gouldii*	Stick-insect	*Phasmidae sp.*
Kangaroo, Red	*Megaleia rufa*	Tortoise	*Emydura sp.*

Apple Box	*Angophora intermedia*	River Red Gum	*Eucalyptus comaldulensis*
Beefwood	*Grevillea striata*	Mallee	*Eucalyptus dumosa*
Belar	*Casuarina cristata*	Stringbark	*Eucalyptus acmenioides*
Bloodwood	*Eucalyptus gummifera*	Paper-bark	*Melaleuca leucadendron*
Lemon-scented Gum	*Eucalyptus citriodora*	Sandalwood (Sugarwood)	*Myoporum platycarpum*
Coolabah	*Eucalyptus microtheca*		

REFERENCES

These refer to the sources of specific information and are numbered according to the small numbers in the text.

1   Sturt, C. (1849), **Narrative of an Expedition into Central Australia.** T. & W. Boone, London.

2   Wills, W. (1861), **The Burke and Wills Exploring Expedition** (From Wills' journals and letters). Wilson and Mackinnon, Melbourne.

3   Jackson, S. W. (1919), 'Haunts of the Letter-winged Kite', **Emu**, Vol. 18.

4   Hollands, D. L. G. (1979), 'The Letter-winged Kite Irruption of 1976-77', **Australian Bird Watcher,** Vol. 8.

5   McGilp. J. N. (1934), 'Hawks of South Australia', **The South Australian Ornithologist.**

6   Cameron, A. C. (1974), 'Nesting of the Letter-winged Kite in Western Queensland', **Sun Bird.**

7   Campbell, A. J. (1901), **Nests and Eggs of Australian Birds.** The author, Sheffield.

8   North, A. J. (1913-14), **Nests and Eggs of Birds found Breeding in Australia and Tasmania.** Australian Museum, Sydney.

9   Cameron, A. C. (1975) 'Nesting of the Square-tailed Kite', **Sun Bird.**

10 Newton, I. (1979), **Population Ecology of Raptors.** Buteo Books, Vermillion.

11 Brown, L. H. and D. Amadon (1968), **Eagles, Hawks and Falcons of the World.** Country Life Books, London.

12 Serventy, D. L. and H. M. Whittell (1976), **Birds of Western Australia**, 5th Edition. University of Western Australia Press, Perth.

13 Barnard, H. G. (1914), 'Northern Territory Birds', **Emu**, Vol 14.

BIBLIOGRAPHY

The following publications contain useful information on Australia's birds of prey.

Brown, L. H. and D. Amadon (1968), **Eagles, Hawks and Falcons of the World.** Country Life Books, London.

Campbell, A. J. (1901), **Nests and Eggs of Australian Birds.** The Author, Sheffield.

Cayley, N. W. (1968), **What Bird is That?** (5th Edition). Angus and Robertson Ltd., Sydney.

Condon, H. T. (1967), **Field Guide to the Hawks of Australia.** (4th Edition). Bird Observers Club, Melbourne.

Frith, H. J. (Ed.), (1969), **Birds in the Australian High Country.** A. H. & A. W. Reed, Sydney.

Frith, H. J. (Ed.), (1977), **Readers Digest Complete Book of Australian Birds.** Readers Digest Services, Sydney.

Macdonald, J. D. (1973), **Birds of Australia.** A. H. & A. W. Reed, Sydney.

Morris, F. T. (1976), **Birds of Prey of Australia, A Field Guide.** Lansdowne, Melbourne.

North, A. J. (1913-14), **Nests and Eggs of Birds found Breeding in Australia and Tasmania.** Australian Museum, Sydney.

Pizzey, G. and R. Doyle, (1980), **A Field Guide to the Birds of Australia.** Collins, Sydney.

Serventy, D. L. and H. M. Whittell, (1976), **Birds of Western Australia.** 5th Edition. University of Western Australia Press, Perth.

Sharland, M. (1958), **Tasmanian Birds.** Angus & Robertson, Sydney.

Slater, P. (1970), **A Field Guide to Australian Birds Non-passerines.** Rigby Ltd., Adelaide.

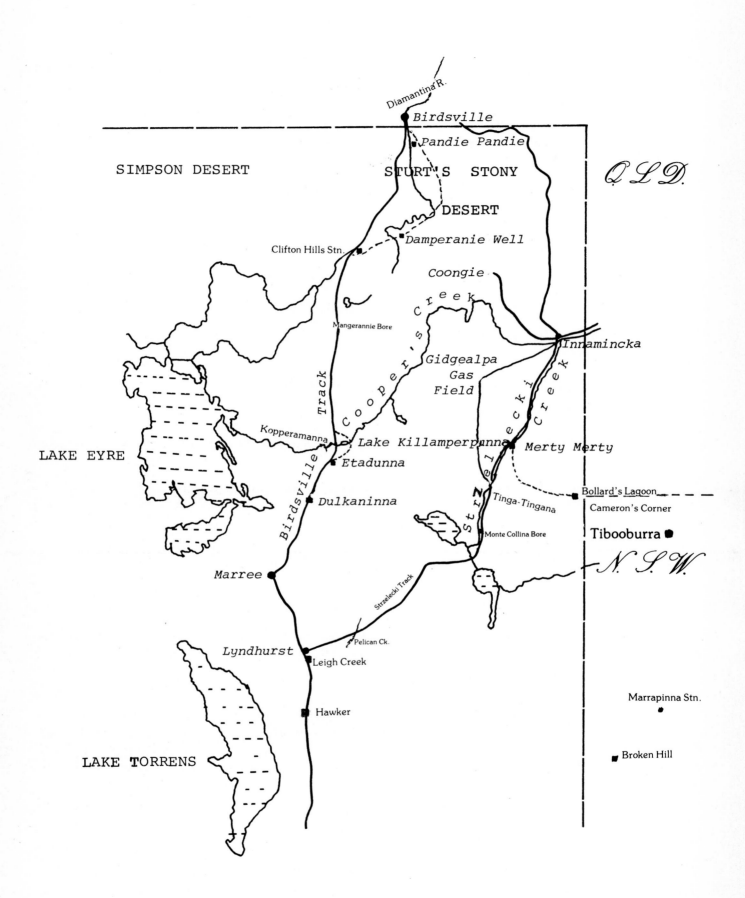

SIMPSON DESERT

STURT'S STONY

DESERT

QLD.

Diamantina R.

Birdsville

Pandie Pandie

Damperanie Well

Clifton Hills Stn.

Coongie

Cooper's Creek

Mangerannie Bore

Gidgealpa
Gas
Field

Innamincka

LAKE EYRE

Kopperamanna

Birdsville Track

Lake Killamperpunna

Etadunna

Strzelecki Creek

Merty Merty

Dulkaninna

Tinga-Tingana

Bollard's Lagoon

Cameron's Corner

Monte Collina Bore

Tibooburra

N.S.W.

Marree

Strzelecki Track

Lyndhurst

Pelican Ck.

Leigh Creek

Marrapinna Stn.

Hawker

LAKE TORRENS

Broken Hill

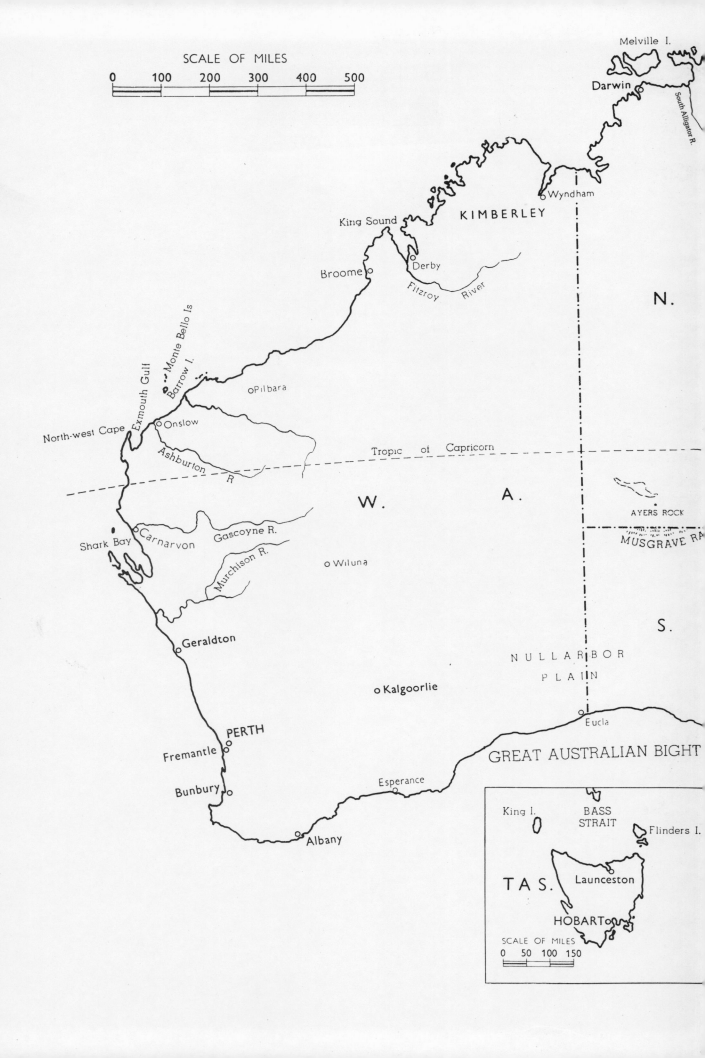